THE OPEN FIELD SYSTEM AND BEYOND

THE OPEN FIELD SYSTEM AND BEYOND

A property rights analysis of an economic institution

CARL J. DAHLMAN

ASSISTANT PROFESSOR, DEPARTMENT OF ECONOMICS
UNIVERSITY OF WISCONSIN, MADISON

CAMBRIDGE UNIVERSITY PRESS

CAMBRIDGE
LONDON   NEW YORK   NEW ROCHELLE
MELBOURNE   SYDNEY

Published by the Press Syndicate of the University of Cambridge
The Pitt Building, Trumpington Street, Cambridge CB2 1RP
32 East 57th Street, New York, NY 10022, USA
296 Beaconsfield Parade, Middle Park, Melbourne 3206, Australia

First published 1980

Printed in Great Britain by The Anchor Press Ltd
and bound by Wm Brendon & Son Ltd
both of Tiptree, Essex

Library of Congress cataloguing in publication data
Dahlman, Carl Johan, 1942–
The open field system and beyond.
A revision of the author's thesis, University of
California, Los Angeles, 1976, issued under title:
The economics of scattered strips, open fields, and
enclosures.
Bibliography: p.
Includes index.
1. Commons–England–History. 2. Inclosures–
History. 3. Land tenure–England–History. I. Title.
HD604.D33 1980   333.2   79 7658
ISBN 0 521 22881 6

CONTENTS

PREFACE

This is an expanded and rewritten version of my doctoral dissertation at UCLA. In retrospect, it is curious how much it is a brainchild of the general intellectual atmosphere at the time of writing and my attending classes. When I first took an interest in the problems of how institutional arrangements deal with transaction and information cost problems it seemed that almost everyone there, without any explicit agreement or understanding, was working on similar or related issues. If there is anything of merit in the following, it should be looked at as a tribute to my teachers – especially A. A. Alchian, H. Demsetz, J. Hirshleifer, A. Leijonhufvud, and E. Thompson. It was an exciting time to be a student there, and I am grateful for the lasting influences that these people have had on my thinking.

My greatest debt is to A. Leijonhufvud. He originally aroused my interest in the open field system in his lectures on economic history, and contributed to the shaping of my final, or at least current, ideas on the topic. Without his continued support, this study might never have been published.

Several other people have read and commented on this work. I am especially grateful to Professor E. Searle for putting up with an economist's sometimes appalling naiveté with respect to historical reality and the work of historians. The interest shown in my work by Professors J. Hirshleifer and D. Shetler was a constant spur to me, and I am greatly indebted to them. I also wish to acknowledge the helpful comments given by D. Bromley, S. Ferris, A. Hyman, E. Jones, R. Kormendi, and D. McCloskey. A special expression of gratitude is due to D. North who as a reader for the Press gave a generous and penetrating evaluation which led to a redrafting of certain portions. As a result, the work is considerably better. To all these people who have shown interest

vii

in my work I gratefully extend my complete exoneration for any errors, ambiguities, and oversimplifications that remain.

Financial support was given me by the American Scandinavian Foundation. I am deeply indebted to those Americans of Scandinavian descent who through their contributions make it possible for students like myself to benefit from the enriching experience of going through American graduate school. I gratefully dedicate this work to them.

Finally, my wife Celia. She has lived with this book almost as long as I have. An economist herself, she has shown greater appreciation for the new ideas contained herein than I myself have been able to at times.

C.J.D.

1

INTRODUCTION

In recent years, economists have increasingly turned to inquire into the foundations of economic institutions. Economic agents are thought of as 'teams' or 'clubs', setting up voluntary organizations for cooperation and mutual gain. There is the beginnings of an economic theory of government and the State, as creations within the economic system. State and local governments are viewed as providers of services demanded by consumers; communication and control in economic organizations receive much attention, and the firm itself is increasingly the object of study as an entity whose existence must be explained. The economics of law and property rights is one of the most rapidly expanding branches of modern economics.

In this developing literature on the 'economics of institutions', however, contemporary economic institutions are so far receiving practically all the attention; and perhaps this is only natural. It is in attempting to understand our current environment that economics, as a social science, can make a claim to contribute to the shaping of human endeavors. While no one would deny the value of 'lessons from the past', to many it still seems more important to deal with immediate problems before attention is shifted to historical ones.

The prevalence of such attitudes notwithstanding, the present study deals with a social and economic institution that has long since disappeared: the open field system in England. The contention will be that there is indeed much for economic theorists to learn from a study of past economic institutions. Historical studies afford some unique advantages that ought to be exploited to a greater degree than economists currently seem prone to admit. The hindsight offered by historical evolution allows us to look at a general problem area from the special viewpoint of the past, and to gain thereby a fresh perspective on many contemporary issues. When viewed in proper analytical perspective, the problems of production, allocation, and income distribution which are encountered in the open field system take on a shape that will look very familiar to the modern eye, but will also provide additional insights into the nature of institutions in general.

Furthermore, historical studies give us something we shall never live to see from contemporary economic institutions: a full life cycle, as it were. We may ask of past institutions why they were first introduced, why they survived for a period of time, and why they were finally replaced by some alternative arrangement. They provide us the unique perspective of knowing what went before, and followed in the wake of, the particular institution we wish to study. In principle, we should be able to determine the function of any institution as an adaptation to an economic problem in a particular environment. What history enables us to do is to define that environment in a more exact manner than is possible in a contemporary setting – we can exploit the advantages of a bird's eye view. Thus, our contention will be that a study of history, far from being a mere virtue, is a necessity in developing a proper theory of economic institutions.

The basic tenet of this study is that the literature on the open field system suffers from an overemphasis on aspects of technology of agriculture and on local and individual characteristics of various open field villages. Lamentably, there is virtually nothing in the literature that attempts to focus on the system as an organization designed to deal with some specific tasks. Yet it is clear that the conceptual problems of the functioning of the open field system as an organizational structure with a well defined purpose represent a degree of importance quite on a par with discussions of technology, records of bylaws, the course of enclosure, and so on. Consequently, a reorientation is attempted in this study: we shall concern ourselves with the open field system as an institution set up for a specific purpose, and attempt to discover what that purpose was. In so doing, we shall find that what have appeared to many as outstanding anomalies of the system, notably the scattering of the strips and the communal ownership of some of the land, will now appear as intelligent solutions that are part of a delicate and complex structure.

The method we employ is that of the so-called property rights paradigm. The essence of this developing branch of economic theory is that ownership and decisionmaking rights are economic choice variables. Rather than being imposed on an economic system from the outside, they are developments within that system, and are designed to fulfill very specific economic functions. One of the crucial aspects of property rights is their function as a mechanism for social control of individual behavior. Thus, property rights constitute a special steering mechanism which induces economic agents to behave in certain ways, and to avoid behaving in certain others. In the emerging literature on property rights, economists study the efficiency characteristics of various kinds of property rights systems. It can be shown that, depending on the particular context, some types of incentive system yield a better economic result than others. At the heart of such demonstrations stands the concept of transaction costs. In economics it is increasingly being realized that exchanges and transactions are not costless, and that many phenomena can only be accounted for by explicitly stating the nature of such costs. The tenet is that, by influencing incentives, property rights can be used to reduce or avoid such costs of transacting, if they are designed and enforced properly. We shall use this paradigm to see if we can cast light on the special ownership and decisionmaking arrangements in the open field system.

It is hoped, therefore, that the present study will provide a fresh view not only of the open field system, but of the general economic problem of collective ownership and control, as well as of the nature and role of transaction costs. This is the ultimate rationale for undertaking this work. If the approach can be shown to be valid, the door is opened to many fascinating questions – questions that ought to interest contemporary theorists as empirical material for testing propositions about economic institutions or organizations.

This study consequently attempts a twofold contribution. First, it aims to bring the recently developed theory of economic institutions to bear on the open field system and the enclosure movement. Second, it draws on this analysis of an historical problem in order to examine critically the current theory of communal property rights and collective decisionmaking agreements,

and to show how property rights theory can be made an exciting
and useful tool for the economic historian.

The question of efficiency

In most of the vast historical literature on the open field system,
the various issues have traditionally been treated separately. The
puzzle of the scattering of one man's land in the open field has
received much attention, and until recently the contention has
been that it was a clearly inefficient element of the open field
system. Several hypotheses can provide alternative versions for
the introduction of fragmentation, but none can explain the persistence of this costly way of holding land. In the same vein, a
traditional explanation for enclosure has been that the open fields
were so inefficient that they would not allow farmers to introduce
new techniques, owing to the vagaries of collective decisionmaking, and that progressive farming necessitated individual
decisionmakers, unhampered by conservative and inefficient
neighbors. Complementary to, and often inextricably woven into,
this dominant theme, there is an important strand of Marxian
influence. The open field system is portrayed as a system of exploitation of working peasants, and enclosure is interpreted as an
intensification of this abuse. Both explanations have a common
starting point : the unanalyzed presupposition that the open field
system must have been economically inefficient, either in the
allocation of resources and the responsiveness to change, or in
the exploitation of the peasants. Hence the open field system
gave way to a more efficient mode either of production or of
exploitation.

This focus on various inefficiencies is clearly unsatisfactory :
why did the open field system ever exist if it was so inefficient?
It becomes impossible to account for the predominance of the
system across Europe and why it persisted over centuries. A theory
founded on a supposition of dumb peasants or dumb landlords
who suddenly realized the inefficiencies of their behavior also runs
into difficulties in explaining the precise timing of enclosure : why
did enclosure not occur some centuries earlier?

A radically different approach is therefore called for. A
methodological angle completely at variance with the traditional

literature will perhaps yield more fruitful insights: only if we postulate that the open field system in some sense represented an *efficient* adaptation to a set of economic problems can we account for the many centuries of its predominance. In such an approach, we shall be able to deal with the open field system as a unified structure, consisting of several distinct elements that require simultaneous explanation. So we must look for a theory that can account for (a) private ownership of the arable, but collective ownership of grazing grounds, (b) individual decisionmaking with respect to many aspects of cropping, but well defined limitations imposed on this decisionmaking by social rules and regulations, (c) the striking scattering of ownership, and its antithesis, consolidation in the enclosure of the village, (d) the reasons for enclosure and its timing, and (e) the origins of the system. The binding constraint on such a theory must be that it must explain why the open field system was, in some sense, a *desirable* economic organization.

The work of some other economists has already cast doubt on the contentions of wholesale inefficiencies in the open field system.[1] However, these approaches have taken a very limited view of the system: they focus exclusively on only one of the aspects characteristic of the open field village. The problem with such a piecemeal approach is that it can only lay one ghost to rest at a time, and cannot therefore establish the contrary presupposition of the overall efficiency of the system. Consequently, we shall insist that any acceptable theory of the open field system must account for all elements simultaneously – it is simply not sufficient to treat the system as a series of separable features.

There is one apparently formidable obstacle to the acceptance of the basic postulate of the open field system as a desirable economic organization. From the standpoint of economic theory, the received doctrine is very clear: communal ownership and decisionmaking are inherently inefficient. In general, such arrangements are associated with overutilization of scarce resources and complications with respect to non-paying consumers. In order to make the case for the open field system as an

---

[1] The work referred to is that of D. N. McCloskey. For particulars of his published work, as well as some comments, see below, especially Chapters 2 and 4.

efficient organization, it is therefore necessary to reexamine
critically the basis of received theory with its general presumption
in favor of private property rights and individual decisionmaking.
Our thesis will be that general economic theory does not imply
the universal inefficiency of communal ownership and collective
control. On the contrary, correctly applied economic theory will
predict that, under certain conditions with respect to transactions
and decision costs, such arrangements will be superior to private
ownership and individual control.

In explaining the particular structure of the open field village
– private and communal rights, scattering, decisionmaking rules,
etc. – as an efficient adaptation to a production problem with
specific information and decision costs, we blend economic theory
and history. On the one hand, it is shown how correct economic
theory can be applied to gain insight into a problem that has
puzzled non-economists for a long time : our theory will account
for many aspects of the open field system and the enclosure movement that hitherto have not been satisfactorily explained. On the
other hand, we are able to draw on the lessons learned from the
study of this historical problem to suggest extensions and correc-
tions of received theory in the area of communal rights and
collective decisionmaking. We will aim to provide a theory that
does *not* rely on an argument that collective 'needs' sometimes
will 'override' private interests : our theory will show how collective property rights and decisionmaking can be quite consistent
with *private* wealth maximization. Thus, we can claim that our
study of a past institution has enriched contemporary economic
theory.

Outline of the study

Before the attempt at formal analysis of the institution of the
open field system, we should outline that system in a manner
which lends itself to economic analysis. Chapter 2 gives a simplified description – timeless and highly stylized – of the institutional arrangements in the open field villages. Also, in order to
establish the correct way of approaching the problem, the various
explanations that have so far been advanced are critically
examined. The theoretical tools to be applied to the problem are

summarized in Chapter 3 : the concepts of property rights and transaction costs, as these relate to the choice of ownership structures as a means of achieving efficient resource allocation. Our theory of the open field system is presented in Chapters 4 and 5. The starting point is a universal feature of the open field system. All open field villages practiced rotating husbandry – the derivation of income by farmers from both livestock and crops in the arable fields. The scale of operation was radically different in these two activities. Each farmer tilled his strips, and often his portion of the open field was very small. Yet grazing was organized on the basis of the whole village : the cattle and the sheep were kept together in large herds, and grazed on extensive areas, on the whole open field or on the commons.

This particular feature of the conditions of production in the open field system can be explained by postulating that livestock production was subject to important economies of scale. The collection of strips into large fields attains the benefits of extensive grazing. On the other hand, the plots in the arable fields were very small – the strips could be just fractions of an acre. An important feature of the production problem of the open field system is, therefore, that the optimal scale of the two major activities was so different.

This interrelationship between the two most important outputs in the open field system provides the key to why collective property rights were preferred to private ownership of the communal grazing grounds. For if the grazing grounds were owned privately, the large-scale grazing areas desired could only be attained by continual transaction between the farmers involved : collective ownership completely bypasses this problem. Furthermore, even with private ownership there would be an incentive for each farmer to over-use the resources belonging to the other farmers, just as there is the problem of over-use of communal property. In addition, with communal property, no one farmer can withdraw any part of the land : with economies of scale in grazing, the threat of such withdrawal is a threat of imposing a negative externality on his neighbors by any one peasant.

In the same manner we can now explain the scattering of the strips in the arable. Here, there is the difficult problem of alternating the same plot of land between two separate activities in which the optimal size of the land is significantly different. There is the additional complexity involved in combining privately owned strips with collective, large-scale grazing. Scattering, as opposed to consolidated holdings, is a means of assuring each farmer that the economies of scale in grazing can be exploited. On the one hand, scattering makes it very costly for any farmer to break away from the others and graze his animals separately for he would thereby lose the benefits of large-scale grazing. On the other hand, scattering will make it impossible for any one farmer to increase his economic gain by threatening to withdraw and induce additional payments from the other farmers, for such a threat would no longer be credible.

This leads to the formulation of some hypotheses concerning enclosure. It is clear that if a technological change occurred to make the optimal size of the arable and the grazing areas equal, the organizations of the open field village would no longer serve any purpose. In such a case, each farmer could easily go his separate way, and the village would break down. However, apart from the difficulty of perceiving such a technological change, the fact that enclosure took hundreds of years to become fully established would be inexplicable. A theory of enclosure based on technological change alone cannot, therefore, be substantiated by the historical facts.

On the other hand, it is easy to show that if an open field village, faced with changing market conditions, would find it economically advantageous to specialize in the production of one of the two major outputs, then the problems of alternating the same plot of land between different activities with divergent returns to scale would no longer be relevant. That is to say, a simple theory of market expansion precipitating specialization in production can accommodate a radical change in the organization of an open field village. At various points in time, it can then be shown how demand, market growth, and technological change caused the precipitation of enclosure in a way that is consistent with both the theory of the open field system developed in this study, as well as with widely accepted accounts of the enclosure movements. In this way we can explain the slow diffusion of enclosure, as well as the concentrated outbursts at different times.

This study will show how the relevant features of the open

field system constituted a superior solution to the alternative of private ownership and individual decisionmaking from the standpoint of both joint and individual wealth maximization, given certain empirically observable conditions. It is therefore possible to construct a unified theory of the open field system that explains all its crucial features as part of the same problem. Therefore, a theory with falsifiable empirical implications that is consistent with the introduction, proliferation, and eventual dissolution of the open field system will be provided. However, it is perhaps not so much in the stress on efficiency or in the use of the property rights and transaction costs approach that this essay departs from previous inquiry into the open field system. Rather, it is in the continued insistence that all features of the system are interrelated and require simultaneous explanation. It is perhaps more important that the validity of this point should be recognized than that the model of the open field system expounded here should become the accepted explanation for the system.

Should the enigma of the open field system ever be satisfactorily resolved, it is likely that we should thereby learn something about the nature of economic institutions in general; for the open field system is simply an instance of how the economic system goes about devising and employing organizations and other institutional arrangements in order to deal with the frictions arising from the interaction of many economic agents. We shall show, in Chapter 6, how the specific model of the open field system elaborated on here can yield additional insights into the functioning of a modern corporation. Further, our treatment of the open field system will offer some glimpses of the nature of economic institutions in general. In the concepts of property rights and transaction costs, we can find perfectly acceptable analytical reasons for why the question of efficient institutions and institutional change can never become a mere question of economic efficiency, in its traditional interpretation, but must firmly be put in the socio-historical context within which any particular institution is designed. Thus, we hope to establish not only that economic theory, properly applied, can be a useful tool in understanding historical institutions, but also that a study of history is a *sine qua non* for any economic theoretician interested in the nature of economic institutions and their change. It is hoped that

this study will also help to remove some of the artificial barriers
that have increasingly been erected between the social sciences.

A note on methods

Apart from employing an only recently developed tool of
economics, this study is perhaps apart from mainstream studies
of the open field system in at least one other respect. No new
empirical material is presented, and no primary sources have been
examined in order to add to the information about the open field
system. We shall rely exclusively on empirical information and
secondary sources already published. Since our intention is to
provide a logically consistent version of the manner in which the
several components of the open field system lock together to form
a pattern with a purpose, it will be sufficient to rely on existing
information. At this stage of research on the open field system,
what is needed is not more figures or more detailed accounts of
various villages, but a coherent framework in which all the dis-
parate aspects of the system fit together.

This does not imply that this study is void of empirical content.
It means only that the empirical material consists of knowledge
that has already been assembled and made available. The level
of empirics is of a different order from that which has become
common in the modern cliometric school. One major reason for
this is that much of what we shall examine is not quantifiable in
ways that are amenable to digestion by computers, and therefore
the application of econometric or statistical techniques is not
feasible. We shall, to a large degree, be interested in transaction
costs under various institutional structures. Again, no methods
have been, or could possibly be, developed for estimating transaction costs under different kinds of institutions. This is because we
do not have access to experimental designs that could allow for
identifying such costs, on the one hand, and few data points from
the observable world that can be used for that purpose, on the
other. The slow pace of institutional change is the ultimate reason
for this.

Yet this study claims to be more than just the development of
a theory. Even though statistical techniques are not applicable
and some variables not measurable, there are more roundabout

ways in which empirical implications derived from the theory can be tested, if only incompletely. We shall argue that if the insights provided by the framework to be developed are correct, then we shall also expect to see similar solutions to similar problems in other contexts. This constitutes a more powerful test of the relevance of the theory than a reader is commonly afforded in the popular cliometric literature. For in spite of methodological sermons given to students of econometrics, the usual procedure of statistically inclined economic historians is to estimate a model on data, and then content themselves with reporting confidence intervals for various parameters. This is not a testing of implications : such testing can in principle only be done by assembling a new and independent sample of data points, on which the results of the first estimation are tested. Since this is practically never done, even by serious econometric historians or econometricians, all that the statistical techniques can provide is a guarantee that the estimated model fits the data. Since it is common to juggle various specifications along with the data themselves until 'reasonable' fits have been obtained, it is curious that the practitioners of this peculiar art of self-delusion claim so much for their method.[2] We shall claim in this study that the insights provided by our theory for the open field system afford some important further insights into phenomena that have nothing to do with the rise and fall of the open fields. Thus, the empirical test is qualitative and roundabout, but nonetheless methodologically sound.

In the final analysis, no empirical study is ever better than the theory that explains the phenomena to be studied. For without a theory, it is neither feasible to decide what data are relevant for an understanding of the problem, nor to establish testable hypotheses about the relationships involved. We now have a wealth

---

[2] For example, one instance may be found in the evolution of the results presented by one cliometric student in the open field system itself. In presenting a theory of scattering, this researcher initially claimed that a certain number of strips were commonly observed, and advanced a theory that was consistent with that number. Later, in view of further data, he found that the 'true' number of strip was 25 percent smaller than his initial estimate, but showed that this was still consistent with his model. That is a rather flexible model, and one wonders if any observations will ever be found that contradict it.

of data on the open field system, as well as several different explanations of various parts of the system. However, there is hitherto no explanation for the totality of relations that formed the open field system. Without such a theory, no empirical estimation can be justified. In this study, we shall provide one such theory.

Even if this study departs from the practices of the modern cliometric school with respect to the use of econometric tools, it adheres closely to the other fundamental methodological precept of the 'New Economic History' (if indeed it is new), in the use of the principles of constrained optimization in deriving propositions about observed behavior. Throughout, we shall postulate private wealth maximization as the fundamental assumption about human behavior. In a study that purports to deal with the behavior of medieval peasants, this may appear to be a daring assumption, and many historians may balk at such a seemingly absurd proposition. Some simple preliminary observations on methodology are perhaps required, therefore.

In many accounts of the demise of the open field system in the enclosure movements, the fundamental explanation for the switch to modern methods of farming – i.e., the abolishment of communal ownership and decisionmaking and its substitution with our modern system of one farmer, one decisionmaker, one owner – usually centers around a change in attitudes or behavior. There is an implicit assumption in these explanations that farmers in the open field system did not primarily have profit maximization as a goal for their behavior, but with the advent of better commercial opportunities or changes in technological conditions, profit maximization became more important, and modern farming methods made their entry. Consequently, it may appear to historians who believe that not only enclosure of open fields, but indeed the Industrial Revolution as well, were due to the 'rise of a commercial spirit', that this book has fundamental methodological flaws. For a basic assumption throughout will be that there was no change in human behavior itself, but only in the conditions beyond human control that set the scene for the decisions and choices made by the farmers of the open field villages.

However, the postulate of profit or wealth maximization as descriptive of human behavior should not be misunderstood. It is not intended to correspond to a textbook definition of how people behave when they have perfect knowledge of all consequences of their choices – on the contrary, we shall show explicitly how it is possible to understand the totality of relationships that formed the open field system only by realizing that knowledge of the consequences of economic activities was imperfect. That is simply to say that we shall include imperfect information in the constraints that set the limits for the decisions taken in the open field villages.

Hence, we shall not argue that the precise *dicta* of undergraduate classroom instruction are necessarily applicable in explaining the open field system. However, we shall argue that the methodology of basic economics is a better way of reaching an acceptable explanation for the puzzles of the open field system, and indeed for historical problems in general : it is a poor theory of the open field system and the enclosure movements that remarks on a behavior change without showing clearly what constraints were altered.

The flaw is not that a theory based on behavioral changes lacks internal logic – it does not. Rather, it is that such a theory is not empirically falsifiable with current measurement techniques. It may be that we shall see the day when it becomes possible to develop unambiguous indices of human behavior and attitudes, but at present it is certain that we do not even know how to begin to solve the problems involved in such a project. There is therefore no possible method of either showing unambiguously that attitudes or behavior did not change, or that they actually underwent a fundamental change. There is no observable historical fact that will decide that issue for us one way or the other. It follows that it is a poor research methodology to *start* by postulating that the demise of the open field system can only be understood as a change in behavior or attitudes : the reason is that *any* change, historical or otherwise, can always be explained by postulating a change in behavior, and we shall never be the wiser for it since the proposition is not falsifiable. As a principle of historical research, this is a dead end.

A superior methodology is to assume that behavior remains constant, but that there are changes in conditions beyond the control of the decisionmakers in the open field villages. From a well specified theory, we may derive predictions about how this would alter the economic activities of the cultivators whose choices we study. Since the constraints that will be relevant have to do with markets and relative prices, crop rotations and other aspects of technology, institutional and legal frameworks, political and other decisionmaking mechanisms, it is obvious that the issues revolve around observable phenomena. Hence, if we develop a theory of enclosure that argues that there were important changes in technology that led intelligent and prudent farmers to enclose open fields, we have derived a theory that is testable empirically : all we have to do is show that enclosed farms employ the new relevant methods, whereas the farmers who chose to remain within the open field system did not have access, for one reason or another, to those same technologies. That is, a theory about changes in constraints deals with observable entities, as opposed to one that deals with changes in behavior. Far from being a dead end, this is a methodology that makes it incumbent on us to undertake proper historical research; to find out what the relevant historical constraints were, and see if these constraints would make intelligent and goal-oriented decisionmakers undertake the activities that we observe as historians.

This does not rule out the possibility of changes in behavior occurring simultaneously with changes in constraint. This is left an open question. Often we shall find that it is not possible to explain all observed historical changes by invoking changes in constraints alone : there will remain a residual that is unexplained. In such a situation, two solutions are conceptually possible. First, we may admit that our measurement techniques are imperfect, and that there are constraints that are relevant but not observable or currently measurable. This simply tells us to go back and do more research. Secondly, however, we might argue that we have all the relevant data of all the relevant constraints, and we are still not able to explain all the observed variations in the historical process. We may then reach the conclusion that since we cannot explain it in any other manner, it must have been due to a change in behavior. That is to say, an explanation for changes in historical processes that relies on changes in attitudes and behavior ought to be a *last* resort : when all other historically observable avenues have been explored and found unsuccessful, then perhaps we shall be forced to accept the unpalatable and unfalsifiable proposition that behavior, after all, underwent a significant change.

It follows that if we believe that the behavior of farmers on enclosed farms is best approximated by an assumption of rational behavior striving towards efficiency, then we shall have to accept the provisional assumption that behavior in unenclosed townships also was rational and intelligent, until proven otherwise. We shall then attempt to find out exactly what constraints made the open field system an intelligent and rational choice. It should be remembered throughout, however, that the terms 'rational and intelligent choice' and 'private wealth and profit maximization' have really a weak and reasonable interpretation : when the technical language is translated into English, it implies that the farmers of the open field system behaved consistently and predictably, and attempted to wrest as good a living as possible from the limited knowledge and means that they had at their disposal – not necessarily that they followed the latest principles of modern management science.

In deriving our theory, the method we employ is, therefore, inductive rather than deductive. From the knowledge that economic historians have accumulated about the open field system, we then attempt to infer what conditions made those observations consistent with each other and with economic efficiency. We treat as choice variables institutional arrangements and ownership rights, in contradistinction to most other studies in which these considerations are taken as initially given.

In utilizing such relatively unfashionable methods, this study runs the danger, perhaps, of not being accepted by mainstream economic historians who employ currently more popular techniques. This study is directed therefore not only to students of the open field system, but also to those interested in the theory of property rights and transaction costs. Apart from various minor applications of property rights theory to the analysis of certain laws, this is the first truly intensive study of a particular economic institution from a property rights perspective. It is to be hoped that it is not the last.

2

THEORIES OF THE OPEN FIELD SYSTEM

Behind the innocuous phrase 'the open field system' two complex but related phenomena are hidden, and they are sometimes not easily distinguished. The first is that the system never was an unchanging and monolithic entity : there never was an open field system that had an identical shape all over England through the centuries from its appearance until its replacement with the modern system of farming.[1] It is by now quite clear that the system was not 'devised' as a finished product of human ingenuity and implanted fully grown by rational, conscious decisionmakers.[2] Rather, it grew over time in an unplanned manner : trial and error must have been the method by which the farmer chose to discard or preserve certain solutions to the problems of farming that he faced in dealing with soil, animals, climate, topography, crops, markets, transportation, and all the other conditions he must conform to in making his living. Thus, at any one moment there were open field villages in various stages of evolution : those that had already adopted what we have come to determine as the mainstays of open field farming, and those that were in the process of developing towards that final state. In addition to this

[1] See, e.g., A. R. H. Baker and R. A. Butlin (eds.), *Studies of Field Systems in the British Isles*, Cambridge 1973, especially pp. 619–20, and also the various contributions to that volume; also, Joan Thirsk (ed.), *The Agrarian History of England and Wales*, vol. IV, *1500–1640*, Cambridge, 1967.

[2] See the discussion between Joan Thirsk and J. Z. Titow on the issue of the exact nature of this growth of the system in *Past and Present*, no. 29, 1964, pp. 10–25, no. 32, 1965, pp. 86–102, no. 33, 1966, pp. 142–7. Baker and Butlin, *Studies of Field Systems*, p. 624, laid to rest the ghost invented by Gray that the open field system was brought to England in its fully fledged shape by the Anglo Saxons, and instead develop a model based on population growth. On the other hand, they also point out that 'remodelling', i.e., the implantation of the system from a less developed form, did indeed occur in various villages (pp. 652–3), so the notion is not altogether far fetched.

problem of evolution, there is the added complication that whatever is understood by the term 'open field system', it never looked the same in all regions, even in a fully developed stage. Since a system of production is an adaptation to local problems as well as to a general 'state of the arts', or the technology of production, it is unlikely that there would ever be a single monolithic structure of agricultural organization in a country that has such varied topographical and regional conditions as does England. To many, it therefore appears that local variations constitute the norm rather than the exception.[3]

Viewed in this light, the open field system is nothing but a shimmering mirage, a self-delusion of scientific minds bent on classifying all phenomena into neatly labelled boxes. Indeed, it is perhaps such a reluctance to abstract from the multitude of local variations that accounts for the inability of economic historians to give a consistent, complete explanation for the open field system.[4] Consequently, the second complex phenomenon behind the label is the lack of agreement of what is to be understood by an 'open field system', as different authors disagree to some extent on what constitutes the salient interlocking features of the system. As a corollary, the accepted methodology in dealing with the open field system seems to be to divide it up into independent pieces. It is common to treat scattering as a problem of its own, separate from enclosure, which constitutes an issue by itself; or, to take another example, to deal with technology of production as the main feature of the evolution of the system, quite apart from and independently of such things as legal structures or urbanization

---

[3] As, e.g., M. M. Postan, who says: 'So great were the variations that no student of medieval agriculture would nowadays dare to assemble all the medieval agrarian institutions into a portmanteau model capable of accommodating the whole of England during the whole of the Middle Ages.' M. M. Postan, 'Medieval agrarian society in its prime. 7. England', in M. M. Postan (ed.), *The Cambridge Economic History of Europe*, vol. I, *The Agrarian Life of the Middle Ages*, 2nd ed., 1966, p. 571.

[4] For an extreme version of this attitude, consider the following opinion about the role of historians: 'In our work we are daily faced with the whole richness of human experience in layer on layer of human societies, whose members never behaved by logical standards but acted and reacted with all the contrariness of which humans are so thoroughly capable. We are responsible for *all* this experience, not only the parts that can be analyzed statistically or explained logically.' P. E. Tillinghast, *The Specious Past*, Reading, Mass., 1972, pp. 14–15.

and market growth. This has led to a curious situation. When different authors use the phrase, they only have in mind rather vague and undefined characteristics of a farming system that was in reality a very complex and diverse structure, and at the present state of research into the open field system, studies of local variations seem to be the norm rather than the exception.[5]

This current approach of economic historians clouds the fact that the open field system, after all, really did constitute a system – for if it did not, there is little justification for studying it. The open field system is, or should be, of interest as a system of agricultural production, not as a collection of quaint local customs that can be recorded with delight and held up for admiration. If the open field system really was a system, it has a degree of importance and general applicability that will transcend the variations : we should consequently look for the theme, and disregard all else.

In adopting this alternative approach we must single out those elements that were intrinsic to the system in its fully developed version and thus reach an agreement on what constitutes the empirical material to be explained, and see how far we are justified in calling it a 'system' that requires a special explanation. Having done this, it is possible to proceed without paying much attention to less than fully developed examples of the system, or to those local variations that made it so diverse, for such variations will then be recognized to have a common theme. If we can discern the theme, the variations will acquire a new meaning and can be seen in a new light.

The science of economics has a tradition that is uniquely apt to deal with this issue. This is the Marshallian method of treating simply a representative firm or consumer, rather than dealing with every firm and consumer as a different agent, each of which requires a different explanation – a task left for the management consultant or the psychologist, as the case may be. The Marshal-

---

[5] Perhaps it is this attitude among historians that accounts for the fact that most of the generalizing work in the area nowadays is done by geographers in Britain, as the contributions in Baker and Butlin, *Studies of Field Systems*, where eleven of the twelve contributors are geographers, or the excellent work by J. A. Yelling, *Common Field and Enclosure in England 1450–1850*, London, 1977; and in the United States by economists, as for example Donald McCloskey's work, referred to below.

lian representative firm or consumer is never thought of as a unit that actually exists : indeed, no one firm or consumer may ever behave exactly as the representative one is predicted to behave. For the purposes of analysis, and of explaining and predicting the behavior of a class of agents, rather than each individual one, it is perfectly satisfactory if all agents on average behave as the representative one is supposed to. It is of no consequence if all firms or consumers actually deviate from the description of the representative agent, as long as the representative one actually represents a distribution that can be ascertained with detailed empirical analysis. This method can be easily adapted to the open field system : we can describe a Marshallian representative open field village, as it were, and incorporate into that description all the features that we agree make the open field system a unique and interesting entity. Having done this, we shall have a structure suitable for economic analysis and confrontation with empirical facts.

We shall then characterize a representative village by a number of 'stylized facts' descriptive of the open field system, and the intellectual exercise will be to set up an economic theoretical model that, in a logically consistent fashion, can account for the simultaneous existence of these observed characteristics. In such an approach we shall be more concerned with the internal logic of the broad aspects of the open field system rather than with the variations which historians delight in documenting. It will be seen that, once we have accounted for the basic underlying pattern of the typical open field village, many minor modifications can readily be introduced, which will add nothing or little to the understanding of the overall logic of the structure itself. This will justify the disregard for the temporal and regional differences that everyone agrees were such a prominent feature of the system : we can simply postpone treatment of them, and deal with the important issue at hand, which is to understand the system as a system.

Before, however, we embark on an exploratory journey, it is prudent to pay close attention to existing maps of the territory into which we shall venture, and try, as critically as we can, to determine how reliable they are. It is of course impossible to determine this without some criterion from which we may judge relative reliability. We shall see that in providing such a frame of reference, the Marshallian method offers an advantage: the existing explanations for the various aspects of the open field system can be held against this blueprint of a representative village, and we have therefore a model for judging the relative merits of those, mostly incomplete, maps that previous research has supplied us with.

The purpose of this chapter is thus twofold. First, we shall describe one version of what is to be understood by the open field system – a version that will probably not be very controversial, but instead can be said to constitute the backbone of the system under consideration. Secondly, we shall survey the literature on the topic, and critically evaluate which of the agreed features each proffered explanation adequately represents; and to what extent these alternative versions constitute a logically consistent explanation for the open field system and its various aspects.

The representative village

Bearing in mind that the purpose of this study is primarily theoretical rather than empirical-statistical or cliometric, here follows a list of the outstanding characteristics of the representative open field village that will be puzzled together into a coherent picture. A convenient method is to group these 'stylized facts' in several different categories: the physical structure of a village, its ownership structure, its institutional structure, its technological structure, and finally its evolution over time. Together, these observations will constitute the empirical material to be explained.

Physical structure. In approaching our representative open field village, a modern observer transported backwards in time would most likely find three eye-catching features of the layout of the village, that, taken together, would immediately underline how very different the system was from modern farming practices. First, he would notice a basic division of the land of the village into two different categories: arable and non-arable land. He might notice that the arable was sown with a variety of crops, and that part of it lay fallow. He might observe that the non-arable, although strangely referred to as the waste by the local inhabitants, had economic uses, especially perhaps for the grazing of animals, that made it an important element in the system of production. Secondly, he would notice that the arable was divided into two or more fields; the exact number might vary from village to village, but they were uniformly of very large size : hundreds of acres in each field. Finally, he might note than within each of the arable fields there appeared rather regular, although not uniform, subdivisions : there might have been ridges or balks or other visible boundaries in the fields, so that the field appeared to consist of elongated, narrow strips.

This then is the most basic classification of the land use in the open field village : arable and non-arable, the formation of large fields, and the division of the fields into strips. An acceptable theory of the open field system must then account for (i) the division of the total land area into arable and waste, and for the fact that the latter was not subdivided, (ii) the formation of large open fields that have given the name to the system, and (iii) the reason for the division of the arable into strips, so strangely scattered about.

Ownership structure. The ownership structure of land in a typical open field village was a complex affair : any given plot of land could have several interested parties, all having well defined rights of different kinds in that land. Even in the eighteenth century there remained vestiges of the complexities of feudal land law, and it is not easy to apply modern terminology to the ownership of land in an open field village without encountering pitfalls. In modern economic terminology, 'private property' in land is probably a more far-reaching concept than ownership in fee simple was interpreted to mean in the open field village : even a freeholder would have to pay dues and fines to the lord, if there was one. On the other hand, a tenant was usually empowered to make decisions that we nowadays associate with private ownership. The fundamental difference between a freeholder and a copyhold tenant did perhaps consist not so much in the fines paid to the lord, but in the protection that the common law afforded them. By the letter of the law, the freeholder held by deed and had access to the king's courts in cases of litigation, whereas the copyholder held by copy of the manorial court roll, and had no protection except by the custom of the manor. Except for this difference, which in the later years of the open field system became less important, there was little to distinguish a copyholder from a freeholder. Both of them could rent out the land, short or long term, they could buy more land, they could usually determine who the next tenant would be,[6] and they both derived their private income from the crops that were grown on their strips. For general purposes, there is thus no reason to differentiate between copyholder and freeholder with respect to ownership: they were both empowered to make fundamentally similar decisions with respect to the land, and were subject to the same restrictions that the village set up in the form of bylaws. Since they did control most aspects of the land, it is perfectly reasonable to define their ownership as private property, for the land represented the capital asset from which they derived their income, and the use and transfer of this asset were controlled by private decisionmaking, subject to the same limitations in both cases.[7]

The implication of this view is somewhat radical, however. It means that we shall disregard the differences between the largest freeholder, the lord of the manor, in cases where the township was also a manor, and his tenants. There are two important historical facts that would seem to mitigate against a proposition

[6] See F. Pollock and F. W. Maitland, *The History of English Law*, Cambridge, 1968, vol. I, p. 380; J. A. Raftis, *Tenure and Mobility*, Toronto, 1964, p. 48. A point made by Postan is that the lord did not prevent exchanges of land, since he stood to gain from the fines paid at the exchange. See his introduction to C. N. L. Brooke and M. M. Postan (eds.), *Carte Nativorum*, Oxford, 1960.

[7] In the modern property rights literature, it is recognized that private property is always attenuated through the imposition of restrictive measures on the owner. 'Attenuation . . . will always signify the existence of some degree of restriction on the owner's rights to (i) change the form, place, or substance of an asset, (ii) transfer all rights to an asset to others at a mutually agreed upon price.' E. G. Furubotn and S. Pejovich, 'Property rights and economic theory: a survey of recent literature', *Journal of Economic Literature*, vol. X, no. 4, December 1972, p. 1140. It is because all private property rights are attenuated in this manner that we can allow ourselves to treat the rights of freeholders and copyholders as private rights, for even though they were attenuated in slightly different manners, the similarity between them is so strong, when it comes to the decisions they made with respect to land use, that it is more useful to treat them as similar than as different.

that the relationship between the lord and his tenants is at the heart of what made up the open field system. The first is that there are abundant examples of townships that adhered to the open field system in spite of the fact that they consisted exclusively of free tenants with no lord; hence, the implication is that the open field system can be understood without reference to the differences in status between the members of the village. Secondly, although in the heyday of the feudal period the lords retained strong influence over who the tenant on any piece of land would be, and over the effective rent that the tenant would pay, this influence gradually waned. In the end, the open field system remained, but the rights of the lord had become only to collect certain fees on the land, and not to decide who the tenant would be. The open field system thus withstood even such drastic changes in relations between the lord and his tenants. While the distinction between lord and tenants may be crucial for analyzing the differences between England and the continent, for example, an understanding of the logic of the open field system does not require more than a stress on the aspects of tenure that were common to fee simple and villeinage, i.e., the fact that all tenants had some scope of individual decisionmaking power and that they all were subject to the same communal influence no matter what the title to the land.

On the other hand, there were large portions of the land that belonged to the village that were collectively owned by the village members. These were the commons, the lands used mainly for grazing that lay apart from the arable fields. Contrary to popular belief, the commons were not open for anybody to use; they did not constitute what economists call a non-exclusive resource, but were the joint private property of the members of the village.[8]

---

[8] 'Both in legal theory and as a historical fact, only the partners in the cultivation of the tillage land were entitled to the pasture rights, which were limited to each individual by the size of his arable holdings. Outside this close corporation any persons who turned in stock were trespassers; they encroached, not only on the rights of the owner of the soil, but on the rights of those arable farmers to whom the herbage belonged.' Lord Ernle, *English Farming, Past and Present*, London, 1968, p. 297. 'In the earlier period the word common implies common exclusiveness quite as much as common enjoyment. The value of a common to the commoners consisted precisely in the guarantee given them by custom that no one might use it except holders of tenements which since time out of mind had a right

A farmer who had strips in the arable, either by copyhold or freehold tenure, would usually also have rights in the village commons.[9] However, except in the special cases where the village community granted individual outsiders rights of common, the use of the commons was restricted to the community.[10] Even so, the commons constituted a collectively owned resource, owned by the members of the village.

There is one further feature of the ownership structure in an open field village which seems anomalous from a modern viewpoint: the arable land would revert from individually owned and controlled property to communally owned and controlled property in a well defined cycle. In villages where common of shack was practiced, the arable would be open for communal grazing each year before ploughing and after harvest: no individual farmer could prevent the rest of the community from grazing his lands. The right of common of shack had come to belong to all farmers by custom, and could thus not be denied them. The same was true whenever a field was left fallow: it might be ploughed up once or twice to turn the soil over, but even so there was some grazing to be had from a fallow field. However, from ploughing until after harvest, the land was privately owned by the tenants of the strips: they owned the output, they did the labor, they had responsibility for the strips.

We therefore have the following stylized facts about the ownership structure of the open field village: (i) the village was defined

---

thereto, and that no one might use it to a greater extent than the custom of the manor allowed.' R. H. Tawney, *The Agrarian Problem of the Sixteenth Century*, London, 1912, p. 238. Also, see Bracton on the *Laws and Customs of England*, Cambridge, Mass., 1977, F. 222 and F. 223 where he discusses how the rights of common may be legally lost and how they are regained when one has been disseised.

9   For an excellent discussion of common rights in the open field system, see E. C. K. Gonner, *Common Land and Inclosure*, 2nd ed., London, 1966, ch. 1. The law recognized various kinds of common rights: common of pasture, common of estover (the right to gather firewood), common of turbary (the right to cut peat or turf), common of piscary (the right to fish), and others.

10 Common of pasture was of two general kinds: common appendant and appurtenant. The former is the ancient kind that was tied to tenure of ownership of arable land, the latter was recognized in the statute Quia Emptores and paved the way for the separation of common rights from the holding of arable land. Gonner, *Common Land*, pp. 8–11.

as consisting of a certain amount of land, divided into commons and arable; (ii) the arable, when used for cropping, was privately owned and controlled; (iii) the non-arable land was owned collectively by the village community; (iv) the privately owned arable became collectively owned grazing areas, and then reverted to privately owned arable again.

Institutional structure. With the collective ownership of the commons and the reversion of the arable from private to collective property, there were several kinds of interaction between the members of the village : the changeover from private to collective rights and back again in the arable must be settled, the use of the commons controlled, joint expenses shared, and so on. In villages that were also manors, this was done in a formal court under the supervision of the lord or his steward; in villages that were not manors or in vills that consisted of several villages, a village meeting was the formal decisionmaking body.

The court or the village meeting was thus the institutional organization through which internal joint affairs were dealt with. In order to understand the functioning of a typical open field village, we must thus explain the scope of its decisionmaking powers, and analyze the decisionmaking process through which its authority was established. It is almost certain that no formal democratic process or counting of heads constituted the basis on which decisions were reached. However, this is something about which little is known. In the later eighteenth century some formal voting rules were imposed by Parliament. In order to reach a decision to enclose a village, four-fifths and later three-quarters of the votes in a village were required. This was not in numbers, but reckoned in value of the arable. Hence, peasants with large holdings had greater influence than those with smaller ones.[11] It is most likely that the formal rule instituted by Parliament

[11] In his classic work on village bylaws, Ault says : 'The principal landholders must have been the predominant group in "the community of the vill". Landless labourers would not account for anything, even sharecroppers had little voice. In a village assembly in medieval times "there was almost certainly no counting of heads".' W. O. Ault, *Open-field Farming in Medieval England*, London, 1972, p. 58. Also see J. Blum, 'The internal structure and polity of the European village community from the fifteenth to the nineteenth centuries', *Journal of Modern History*, vol. 43, 1971, pp. 549–52.

approximated to what already was the accepted practice in open field villages, so the collective decisionmaking rule employed can best be understood as voting in relation to size of the economic interest belonging to each farmer in the village.

To cope with the alternative use of the arable in cropping and grazing, the court would have to set ploughing dates and harvesting dates so that the changeover from grazing to cropping could be accomplished in an orderly manner. In practice, this put some limitations on the freedom of planting any crops the farmer wanted, for he could not be out of line with the rest of the village in sowing and harvesting. It is also a matter of historical fact that the court kept its hand in the private exchanges of strips between the village members : implicit in the charging of a fee for transferring the title of ownership to the next holder is the power of the court and the lord to deny some transfers. Finally, the court would control the use of the commons in order to avoid the overuse and underinvestment problems that are usually associated with communal property.

In accounting for the open field village structure, we must therefore explain why (i) certain decisions were taken privately and others collectively, (ii) how the collective decision rule was formulated and what it implied, (iii) the restriction on private exchanges of strips, (iv) the cropping rules that limited private decisions about the use of the arable, and (v) the restrictions on the use of the commons.

Technological structure. There is one broad generalization about open field villages that is striking in view of the great regional differences in climatological conditions and soil qualities in England : except for certain areas in the vicinity of large markets, such as London, or within easy access of cheap transportation (generally along waterways for the time period under consideration) very little specialization in production was prevalent. The open field system is generally associated with a wide variety of ouputs, whereas those areas that did specialize did not have the open field system. In spite of the potential economic gains that might have been obtained from regional specialization in that output for which the local natural conditions were relatively well suited, we invariably find that the open field villages produced both livestock and grain for sale in relatively small localized markets, and that the areas that did specialize in the production of either livestock or grain were located in regions that had access to markets of almost national scale.[12] The key to this would appear to lie in the costs of transportation over land, particularly the transportation of grain. Such costs were simply too high to permit exploitation of potential regional advantages.

There is one additional observation that must be included in the conditions that constitute the production problem. It has already been noted how the typical unit of arable cultivation, the elongated strip in the open fields, was of rather insignificant size – usually, the inference, correct or incorrect, is drawn that the strip represented a day's ploughing. However that may be, the point is clear : the strip was cultivated principally by one peasant, and this determined its size. In contrast, the commons that served as grazing areas for the livestock were typically large areas, and the arable fields which were used for grazing between harvests and during years of fallow were also large in size. It is also to be noted that the herd of cattle or flock of sheep grazing on these lands consisted of the animals belonging to all the members of the village community. We must therefore inquire into what technological conditions of production gave rise to a small unit of cultivation in the arable but to a large unit in the grazing of livestock.

In several stages of the cultivation of the arable, the farmers in the village cooperated so that the unit of cultivation was the whole open field itself, or sometimes a subdivision of it called a furlong. This was true for ploughing and harvesting in particular: for these activities it would not be necessary for the peasant to move from one of his strips to another one further away, and then on again. Rather, the plough team would start in one end of the field, and plough each strip in succession until it reached the other end; and a similar action was true for cutting, reaping, and gleaning. The strip therefore constituted a unit of cultivation only for such activities as sowing and weeding, done on an individual basis, and each farmer was entitled to the crops grown on

---

[12] Gonner, *Common Land*, p. 36, also notes this remarkable state of affairs, and also explains it by transportation costs.

his strips. It is to be noted how the organization of the village permitted the attaining of the benefits inherent in large-scale operations in activities where that was desirable, but kept the strip size small for such activities that did not require larger plots of land.

This is also true in the production of livestock. The village might jointly appoint and employ a shepherd for the flock and a herdsman for the cattle, and the livestock usually grazed together in a large herd on a large area of land, whether it be the commons or the open fields themselves. For the production of livestock, it is consequently most appropriate to regard the village as an entity as the unit of production, in contradistinction to the cropping, where the unit must be considered to be the individual peasant.

We must therefore account for (i) the condition that each open field village, whatever the quality of its soils, was constrained to producing both classes of outputs, livestock and grain, (ii) the differences in the scale of land use in cropping and grazing, and (iii) the nature of the interrelationship between the two classes of outputs and the restrictions on organization that this implies.

Evolution of the system. It is most unlikely that all the elements of the open field system referred to above were introduced simultaneously. It is more probable that each characteristic of the system was introduced as a response to some particular problem that became apparent over time, and that various methods of coping with such a problem were tried and the solution of the open field system crystallized as a result of adaptive behavior over several centuries. There is also the rather peculiar fact that some areas of England, notably in the pastoral regions, seem to have abolished the system at the same time that it appears to have been introduced in part of the north.[13] This would seem to imply that the open field system is more than just an historical anomaly, for otherwise the conscious introduction of it would scarcely occur simultaneously with its abolition in other parts of England.

---

[13] For example, Elliott notes: 'Up to the end of the sixteenth century the area of common fields in Cumberland continued to expand.' G. Elliott, 'Field systems of northwest England', Baker and Butlin, *Studies of Field Systems*, p. 76. This was at a time when the Crown had spent almost a century trying to control enclosure and the conversion of common fields to pasture in other parts of the country.

On the other hand, the system did not disappear in one fell swoop either. Although there were two notable outbursts of enclosure of open field villages, the first in the late fifteenth and beginning of the sixteenth centuries and the second in the last half of the eighteenth and the beginning of the nineteenth centuries, it was a process that was never really at a standstill. It was known from ancient times; and throughout the history of the open field system, enclosure and the abolition of open fields was a widely known if not always heartily welcomed phenomenon.

Thus, an adequate model of the open field system must account for (i) the growth of the system and the introduction of its salient features, (ii) the fact that it disappeared in some areas at the same time that it was introduced in others, (iii) why it was abolished in the enclosure movements, and (iv) why the timing of the enclosure decision seems to vary so much.

Theories of the open field system and the enclosure movements

With the broad outlines of the open field system thus set forth we now turn to the various explanations that have been proffered to account for the evolution and eventual dissolution of that system. In this section we shall summarize and critically evaluate a few of these attempts at explaining the apparent anomalies of the institutional arrangements in the open field villages. It is perhaps fitting to start with the explanations offered by historians. The literature on this topic is vast and diverse, and it is difficult to categorize and classify different authors very precisely. However, it is not impossible to distinguish certain major themes, and one such is the question of the efficiency of the system. In the writings of historians there is sometimes a strong undercurrent of Marxist influence : we shall also deal briefly with this theme. Lastly, some contemporary economists have shown an interest in the problem areas, and we shall offer some remarks on their work as well.

The traditional hypotheses. The first problem to deal with in an explanation of the open field system is the question of its origins. Nothing definite will ever be said on this, however, for the evidence that would allow for an incontrovertible answer does not exist. The scanty documentary evidence that does exist will not

allow for any definite inferences. For example, consider one of
the oldest pieces of evidence adduced to prove the existence of
open fields at an early time : 'If ceorls have a common meadow
or other shareland to enclose, and some have enclosed their share
while others have not, and cattle eat their common crops or grass,
let those to whom the gap is due go to the others who have enclosed their share and make amends to them.'

This is one of the laws preserved from King Ine of Wessex.[14]
It was issued sometime between 668 and 694. The law seems to
attest to the existence of common fields – 'common crops or grass'
– and this would indicate the early existence of open field systems.
Yet there is really nothing in the law to prove the existence of the
other elements of the open field system. There is nothing, for
instance, about strips, cropping rules, common grazing, or communal regulations through the court. All the law really shows is
that ceorls shared in the fencing of meadows or arable in which
they may have grouped individual or communal holdings.

The origin of the open field system is therefore very unclear. It
is not even certain that all the elements referred to above were
introduced simultaneously. The evolution of the system may have
been very gradual, where additional elements were brought in as
they became desirable over time.[15] It is not clear, moreover, just
which of the crucial elements was introduced first. Perhaps it
would be a fair generalization to state that the currently accepted
view among historians to account for the concurrent presence of
all the characteristics of the system would go in the following
manner. The pivotal element seems to be the scattering of the
strips; for this, as will be seen below, several reasons can be
adduced, and they all have in common that scattering was introduced for a good reason, but that it involved certain costs. Some
of these costs could only be coped with through the introduction
of communal regulations. For instance, in order to graze on the
arable during fallow and after harvest, common cropping regu-

---

[14] Quoted by Joan Thirsk, 'The common fields', *Past and Present*, no. 29,
December 1964, p. 9.

[15] This seems to be the view of Continental scholars who have studied the
spread of the system there. See R. C. Hoffman, 'Medieval origins of the
common fields', in W. N. Parker and E. L. Jones, *European Peasants and
Their Markets*, Princeton, 1975. Also, the discussion referred to above by
Baker and Butlin in the postscripts to their book.

lations were necessary. Such practices came naturally, as the strips were small and difficult to graze without impinging on the neighbors' arable, and also because communal ownership was a very old practice among Germanic tribes. Communal ownership, then, is one of the oldest elements in the system; however, this applied only to the waste. The communal controls over the arable are of an altogether later date, and as such controls were most probably introduced in response to population growth, the assembly of cultivators that controlled the exercise of common rights was constituted. The chain would therefore lead from the communal ownership of the waste, through to some arrangement giving rise to scattering, and then to communal grazing and control.

The link in this chain that has attracted most attention is undoubtedly the scattering. It is a fascinating element in the open field system, for it seems so clearly inefficient, and appears therefore as an anomaly that demands explanation. Many explanations have been offered, some of which have been accepted for some period of time, only to be discarded later in favor of some alternative. Among the first to work on the problem was Seebohm. He had harsh words for the open field system :

Now, judged from a modern point of view, it will readily be understood that the open field system, and especially its peculiarity of straggling or scattered ownership, regarded from an agricultural point of view, was absurdly uneconomical. The waste of time in getting about from one part of a farm to another, the uselessness of one owner attempting to clean his own land when it could be sown with thistles from the seed blown from the neighbouring strips of a less careful and thrifty owner; the quarrelling about headlands and rights of way, or paths without right, the constant encroachments of unscrupulous or overbearing holders upon the balks – all this made the system so inconvenient, that Arthur Young, coming across it in France, could hardly keep his temper as he described with what perverse ingenuity it seemed to be contrived as though purposely to make agriculture as awkward and uneconomical as possible.[16]

However, Seebohm proceeds to argue that there must have been a justification for the scattering at an earlier date, although it may have lost force over time. His own explanation argues that there was a close connection between the oxen in a plough team and the holdings of the members of the team. He explains :

[16] F. Seebohm, *The English Village Community*, London, 1912, pp. 15–16.

. . . the normal plough team was, by general consent, of 8 oxen; though some heavier lands required 10 or 12, and sometimes horses in aid of the oxen. (. . .) The hide or carucate seems to be the holding corresponding with the possession of a full plough team of 8 oxen. The half-hide corresponds with the possession of one of the 2 yokes of 4 abreast; the virgate with the possession of a pair of oxen, and the half-virgate or bovate with the possession of a single ox; all having their fixed relations to the full manorial plough team of 8 oxen. (. . .) One point more, however, still remains to be explained before the principle of the open field system can be said to be fully grasped, viz. why the strips of which the hides, virgate, and bovates were composed were scattered in so strange a confusion all over the open fields. (. . .) On the hypothesis . . . that the hides, half-hides, virgates, and bovates were the shares in the results of the ploughing of the village plough teams – in other words, the number of strips allotted to each holder in respect of the oxen contributed by him to the plough team of eight oxen – it is perfectly natural that in a grant of *some* only of the holdings the boundaries given should be those of the whole township, viz. of the whole area, an intermixed share in which constituted the holding. (. . .) So long as the limits of the land were not reached a fresh tenant would rob no one by adding his oxen to the village plough teams, and receiving in regular turn the strips allotted in the ploughing to his oxen. In the working of the system the strips of a new holding would be intermixed with the others by a perfectly natural process.[17]

One of the problems with this explanation, however, is its implication that in areas where the soil did not require the heavy, wheeled *caruca*, but only the smaller *aratrum*, no scattering should be observed. For the smaller plough did not require the full 8-oxen team, so dividing the land according to shares on the team would not yield scattered holdings. Furthermore, an obvious counter-argument to the proposition is that it would be perfectly possible to adjust the rental price of the team of oxen so that the costs of scattering could be avoided. The heavy plough theory of the scattering of the strips must needs argue, therefore, that the costs of transacting in the hiring of the oxen were higher than the costs involved in the scattering of the strips, a proposition that seems somewhat difficult to uphold in view of the fact that the rental price of the oxen must have been subject to negotiation anyway. For there was, of course, no necessary and definite rela-

[17] *Ibid.*, pp. 65–6, 113–14.

tionship between the input of oxen from one man into the team and the amount of land that the team would plough : if he put a greater proportion of oxen into the team than his proportion of the ploughing he would require compensation; in the opposite case he would have to pay some. The plough may have been heavy, but there is no need to assume that it made the peasants exert themselves to the point where they did not care to assert themselves when being taken advantage of.

Instead, the argument was advanced that it was a desire for equality on the part of the peasants in the village that accounted for the scattering. Homans says :

. . . the strips of the different holdings should lie so scattered over the fields that each holding partook in equal proportion of the good and bad soils and locations of the village. (. . .) Under the circumstances, it was wholly natural. Since they did not know the techniques of today, villagers of the thirteenth century had only dung and marl with which to increase the fertility of their fields, and these only in small amounts. They could have no escape from, no way of progressively improving upon the original qualities of their soils. It would be no wonder, then, if they, or their forefathers who originally laid out the villages of the champion country, insisted that the different holdings would share proportionately in the areas of the village in which good and bad soils were found – not to speak of the matters of exposure and drainage – and such sharing would necessarily mean that every holding would consist of a number of parcels scattered over the fields.[18]

Vinogradoff made the same case :

The only adequate explanation of the open-field intermixture . . . [is] the wish to equalize the holdings as to the quality and quantity of land assigned to [all tenants] in spite of all differences in the shape, the position and the value of the soils.[19]

A similar opinion is given by Lord Ernle :

To divide equally the good and bad, well and ill situated soil, the bundle of strips allotted in each of the three fields did not lie together, but was intermixed and scattered.[20]

[18] G. C. Homans, *English Villagers of the Thirteenth Century*, New York, 1941, pp. 90–1.

[19] P. Vinogradoff, *Villainage in England*, Oxford, 1892, p. 231.

[20] Lord Ernle, *English Farming*, p. 25.

Many other quotations to the same effect could be found. It is
an explanation which has not held up, for there were such great
differences between the tenants on any one manor that it is impossible to sustain for very long the illusion that property rights
arrangements were governed by a desire for equality. Instead,
many have argued that scattering arose because of the effects of
partible inheritance, as for example Joan Thirsk :

> As population increased, new households were at first accommodated
> in the old settlements. It was this increase which led to the emerg-
> ence of nucleated villages. Other changes accompanied the growth
> of population. Farms were split up to provide for children, and fields
> were divided again and again. (. . .) As farms were divided into
> smaller units and population rose, the production of food had to be
> increased. The arable land was enlarged by assarts and the fields
> became more numerous. (. . .) . . . as fields multiplied whenever new
> land was taken into cultivation from the waste, . . . the parcels of
> each cultivator became more and more scattered.[21]

The problem with this theory is that if it is followed *ad infi-
nitum*, and there seems to be little reason not to do this since the
open field system existed for perhaps a thousand years, the last
generation should end up with infinitely small pieces of land
scattered in an infinite number of places; while, quite to the contrary, we do know that the size of the strips remained stable,[22]
although the particular local size may have varied from village to
village according to local conditions. Another problem with this
theory is of course the result of marriage between persons from
different villages, for then the scattering would appear as between
villages as well; and that is rarely observed. Indeed, if it had been
a common occurrence, the village structure would soon have
broken down altogether. In addition, the partible inheritance
theory must, of course, show that the costs of bartering or renting
the land were higher than the costs implied in the scattering.[23]
Most important perhaps, it seems that only land held in fee simple

---

[21] Joan Thirsk, 'The common fields', p. 9.

[22] Joan Thirsk, *Agrarian History*, vol. IV, p. 11, points to the existence of a
lower bound to the size of the strips.

[23] This would be a difficult proposition to uphold. 'More important as a
counterbalance to the divisive effects of partible inheritance was the free
alienation of land *inter vivos*. This freedom stimulated the market, and
enabled the more enterprising and more prosperous tenants to augment

was passed on in partible inheritance.[24] This is inconsistent with the fact that the villein tenants' holdings were scattered in the same way as the freeholders'.

An alternative argument relies not solely on partible inheritance, but brings in assarting as an additional element. Bishop has shown that free tenants cooperated in the intake of new land to the village, and that this could lead to scattering if the peasants divided the new lands among themselves, taking it in turns to do so; thus, over relatively long periods of time, scattered holdings would emerge.[25] A new piece of land would naturally be cleared adjoining the existing arable, and would have its designated owner. Each year more and more plots would be added, and after many years the holdings of one family would be scattered all over the field depending on when each parcel was taken into the field. This theory must assert that the costs of exchanging one strip for another by barter was so high that it did not warrant the gains to be had from consolidating the strips. Moreover, the theory must show that there was no adequate way of determining the appropriate rental price of a parcel of land : for if it were possible to determine the rental price, one peasant could save on the costs of scattering by renting the land from some other tenant adjoining him and simultaneously leasing out another parcel in another part of the field, thus achieving consolidation. Yet we know that there was a fairly active market in renting land where the proper rental price of a parcel was determined efficiently and with apparent ease. So the argument based on the intake of new land logically implies an assertion that sometimes the peasants knew how to determine the rental price of land, but that at other times they did not.

Recently, McCloskey has shot a broadside through this fleet of arguments, and it would appear that he has effectively demasted all of them. He says : 'One or other *deus ex machina* – common

---

their holdings by purchases or lease and resulted in a growing inequality in the size of land holdings.' A. R. H. Baker, 'Field systems of southeast England', in Baker and Butlin, *Studies of Field Systems*, p. 409.

[24] M. M. Postan, *The Medieval Economy and Society*, Berkeley and Los Angeles, 1972, p. 146.

[25] T. A. M. Bishop, 'Assarting and the growth of the open fields', *The Economic History Review*, vol. VI, 1935–6, pp. 13–29, *passim*.

ploughing, egalitarian instincts, clearing of the waste, partible inheritance – is lowered into the action to scatter the plots, but when it has been lifted back into the rafters the question arises why its effects persist.'[26]

McCloskey's point is that, even though some or other of the arguments advanced for scattering may account for the *origin* of scattered holdings, it is impossible to understand why scattering persisted if it was as inefficient as Seebohm maintained in the quotation above. The ability to exchange land should have made self-interested peasants transact with each other so that the negative effects of scattering could be overcome. Yet this did not seem to occur: scattering persisted until the enclosure of the whole village put an end to its inefficiencies.

The puzzle of why the open field system with its scattered strips persisted for such a long period of time over so vast an area has therefore not yet been satisfactorily resolved. We know that variations of it existed from the Ural mountains in Russia to the Atlantic Ocean, north of the Alps,[27] for at least a thousand years,[28] but we do not know why. McCloskey, after going through a taxonomy of possible reasons, accepts a theory of scattering based on risk sharing;[29] however, that theory has certain shortcomings as well. For the moment we postpone a discussion of that theory, but we shall return to it.

Historians have traditionally proffered two complementary explanations for enclosure: the increase in the wool trade is made to account for Tudor enclosures, and the Agricultural Revolution for the enclosures in the eighteenth century. Tawney states:

The explanation [for enclosures] most generally given by contemporary observers was that enclosing was due to the increased profitableness of pasture farming, consequent upon the development of the textile industries; and . . . there does not seem any valid reason

26 D. N. McCloskey, 'The persistence of English common fields', in Parker and Jones, *European Peasants*, p. 95.

27 P. Mantoux, *The Industrial Revolution in the Eighteenth Century*, London, 1947, p. 148.

28 Joan Thirsk, 'Field systems of the east midlands', Baker and Butlin, *Studies of Field Systems*, p. 245.

29 D. N. McCloskey, 'English open fields as behavior towards risk, pp. 124–70 in P. Uselding, *Research in Economic History: An Annual Compilation*, vol. I, 1976.

for disputing it. The testimony of observers is very strong; . . . and with scarcely an exception they point to the growth of the woollen trade as the chief motive for enclosing.[30]

Recent evidence has cast doubt on this explanation. Cohen and Weitzman ran a regression on the relative price of wool to grain, and were unable to substantiate the hypothesis :

A major problem with the wool trade argument is that the price data simply do not support the argument. If the analysis were correct, we would expect the price of wool to rise relative to the price of grain. On close inspection of the available data we can find no systematic difference in the trend of wool and grain prices between 1450 and 1550. If anything, the price of wool declines relative to the price of grain. The data so blatantly contradict the standard analysis that it is difficult to understand how it has managed to maintain such general acceptance.[31]

At first, this seems a very important objection to the wool trade explanation. Yet this is peculiar : for the early enclosures most definitely were for pasture, and the manufacture of cloth and the export of woollen products certainly grew at the time. It is perhaps likely that we have here a classic case of what econometricians call the identification problem : it is quite possible that the volume of wool produced increased enough to offset any tendency for the price to increase as demand for wool products increased – that is to say, we have a case of simultaneous shifts of supply and demand schedules. If that is so, there would be no reason to predict any increase in the price of wool, and it would still be possible to uphold the traditional theory that the early enclosures are very closely connected with the wool trade.

On the Agricultural Revolution, Gray states :

. . . agricultural progress in England would ultimately depend on the disappearance of the open-field system. A form of tillage so inconvenient, so inflexible, so negligent of the productivity of the soil, could not long endure after technical improvement in ploughing had made possible its abandonment and after its social advantages had come to be disregarded.[32]

---

[30] Tawney, *The Agrarian Problem*, p. 195.

[31] J. S. Cohen and M. L. Weitzman, 'A Marxian model of enclosures', *Journal of Development Economics*, vol. I, no. 4, February 1975, p. 318.

[32] H. L. Gray, *English Field Systems*, Cambridge, Mass., 1959, p. 109.

Modern writers, as for example Chambers and Mingay, present much the same views :

There were broadly four main objects behind an enclosure, whether by Act or otherwise. First enclosure undoubtedly made for more efficient farming by making farms more compact, larger and easier to work, by making possible a better balance between arable and pasture, by encouraging the adoption of alternate and convertible husbandry, and by allowing better care of animals – in short by overcoming the defects of much of open-field farming. (. . .) A second object of enclosure was to convert land to more profitable uses. (. . .) The enclosure of commons and wastes achieved a third objective, namely that of expanding the areas of land under regular cultivation . . . (. . .) Finally, little-noticed objects of some enclosures were to improve efficiency by getting rid of tithes and by bringing order and greater convenience to a parish. . . .[33]

Again, these are arguments which are very difficult to uphold. The implication is that the open field system was very slow to adapt to new techniques as they were invented, and much modern research has amply shown that this certainly was not the case. The new crops associated with the Agricultural Revolution were introduced in open field villages, the floating of water meadows was arranged through village cooperation, and even selective breeding methods were employed in the champion villages.[34] Too much stress on arguments of this kind produces a difficulty greater than the one 'solved' – namely, to accept the fact that open field villages ever existed at all : if scattering was so inefficient, why was it not abolished as soon as the advantages in the form of equality, ploughing, assarting, or inheritance had already been exploited? It is hard to find a more appropriate label for the traditional explanations than 'the dumb peasant model', for they are predicated on the basic belief that the peasants of the open

[33] J. D. Chambers and G. E. Mingay, *The Agricultural Revolution*, London, 1966, pp. 79–80.

[34] Even Lord Ernle recognized this : 'In 1773, an important Act of Parliament had been passed (13 Geo. III c. 81), which attempted to help open-field farmers in adapting their inconvenient system of occupation to the improved practices of recent agriculture. Three-fourths of the partners in village-farms were empowered, with the consent of the landowner and the tithe-owner, to appoint field-reeves, and through them to regulate and improve the cultivation of the open arable fields.' *English Farming*, p. 224.

field villages were unable to run their affairs in such a way as to achieve a reasonable degree of efficiency.[35]

Nor is this view of the reasons for the persistence of the open field system a thing of the past. In modern literature on the subject a more sophisticated picture is painted, but it is still one that has the 'dumb peasant' view as the underlying logic. We may exemplify this by Yelling's very recent general treatment of the problem of enclosures and the open field system. In dealing with the question as to why some villages stubbornly refused to enclose, Yelling begins by the following precept when discussing the problem in the context of east Leicestershire : '. . . there is little doubt that its [the region's] optimum land use pattern would have approximated to that of the enclosures rather than that of the common fields.'[36]

To begin with, there is a difficulty here of interpreting exactly what is meant by 'optimal land use pattern', for Yelling does not make the distinction between economic and technological efficiency. It is quite possible that land that does not use the best or most recent advances in techniques of production still makes more profit, or yields a higher income, than land that employs the most efficient farming methods from a technical perspective : technical efficiency may or may not imply economic efficiency. However, it may well be that the situation in the region under consideration was such that the two concepts coincide, so that the technically most efficient farming methods were also the most economical ones, in which case Yelling is justified in disregarding the distinction. The point is raised here only to indicate the possibility that, unless the distinction is kept in mind and thoroughly examined using data on prices as well as productivity, cases may occur where open field villages indeed allowed for a greater income with older techniques than enclosed farms did with new

---

[35] Curtler presents a view in this vein : 'The old common field system was only suited for a primitive state of society and was bound to disappear with the advance of civilization. It was extremely wasteful; the scattering of the strips all over the open fields led to an astonishing waste of time, and confusion : the pace of the common work was set by the worst farmer; therefore, no individual initiative or enterprise was possible.' Dumb peasants, indeed. W. H. R. Curtler, *The Enclosure and Redistribution of Our Land*, Oxford, 1920, p. 63.

[36] Yelling, *Common Field and Enclosure*, p. 197.

techniques. In such a situation it is not at all clear that enclosure represented the optimal land use pattern. The point is that the word 'optimal' is open to several interpretations if no point of reference is clearly specified : optimal relative to technology or relative to profits.

Granted, however, that Yelling's precept of the superiority of the enclosed farms over open villages is substantially correct, the enigma of why so many villages remained unenclosed needs to be examined. To account for this, Yelling advances what can be interpreted as three different arguments for why it was sometimes understandable that villages did not enclose when they seemed to have the opportunity to do so. The first argument contends that :

. . . although the great contrast in types of production on common field and enclosed land in the region reflected the fact that the presence of the former did indeed prevent land being put to its best usage, it did not necessarily mean that the management of the common fields was irrational or inefficient given the constraint that this form of agriculture be maintained.[37]

This means that, even if the open field villages were relatively inefficient, their farmers still tried to employ the best farming methods suited to the open fields. Yelling does not at this point specify what he means by the 'constraint that this form of agriculture be maintained'. Several possible interpretations are available, and one version would be to argue that open field husbandry presented its farmers with certain psychological advantages, such as being the traditional way of farming or giving a sense of community, for which they were willing to pay a price in terms of lower efficiency and hence lower income. This pattern of behavior is observed even in modern days, as for example among the Amish sect in North America and among the hippie communes in California. In their refusal to employ modern farming methods, these groups accept a lower income as the price they willingly pay for a different, and in their eyes, superior life style. This does not necessarily make them either irrational or 'dumb peasants' : it can be interpreted as a perfectly intelligent way of reaching a desired goal. However, that same interpretation is not possible in the context of the open field system, simply because so

[37] *Ibid.*, p. 199.

many villages were converted to enclosed farms. This can only be interpreted to mean that the farmers who delayed conversion to the new husbandry changed their mind about paying the price of preserving the intangible advantages of the open fields, and decided to attempt to increase their income instead. The point is that this is a sudden change in behavior, in the light of which the older mode of behavior appears as rather irrational. It is because enclosure inevitably did away with the open field villages that a reliance on the intangible advantages of the open field system necessarily implies the 'dumb peasant' view of the world.

However, this may not be what Yelling has in mind with the word 'constraints', although it is clear that other writers have had that idea in mind, and this is the reason for dwelling on it here. Yelling does advance one alternative interpretation of the word :

We have to interpret such a constraint – and this applies equally to the Midlands – less as an inherent and definite barrier than as a series of practical impediments arising in the course of a continuing political debate. (. . .) It was not simply the result of stubborn conservatism on the part of whole communities, but a logical outcome given the problems posed and the political framework within which decisions had to be made. This argument is admittedly speculative, but it does have the advantage that proprietors in the surviving common-field villages do not have to be regarded as a group whose character and behaviour were in a class totally apart from that of their neighbours. . . .[38]

Again, Yelling does not specify what he means by the 'problems posed and the political framework within which decisions had to be made'. Presumably, he means the difficulties of reaching an agreement, both on the decision to enclose and on the terms of the compensation to be paid. But if that is so, then there must have been different political constraints in those villages that enclosed with relative ease and those that did not accomplish such a conversion to the new system. Only if the constraints were really different can we justify an assertion that it was not a difference in behavior or in outlook that made some villages enclose at a different point in time from others. It is hard to see how such an hypothesis can be upheld, for certainly the decisionmaking rules and the political influence in any one village did not differ much

[38] *Ibid.*, p. 200.

from that of the next, in which case there is no reason to expect
that the political institutions were significantly different in the villages that enclosed from those that did not. If the political constraints were different this must then have been because some
peasant stubbornly refused to go along with the majority, so that
the whole village was prevented from taking what would have
been the optimal course of action. This is not to say that all
farmers were dumb. But perhaps some of them were, and in sufficient strength and number to prevent reason from prevailing.
This is a more sophisticated version of the dumb peasant argument, to be sure, but one that is essentially similar to older views.

There is yet a third argument for the persistence of common
fields in Yelling's text :

. . . the common fields, taken as a whole, were relatively inefficient
even within the broad economic objectives which they set themselves.
However, there is no way of measuring the extent of such a lapse,
for it mainly depends not on features such as the presence or absence
of permanent hedgerows, but on the efficiency of management.[39]

The argument here is that the open fields were not only inefficient relative to enclosed farms, but that they were also used
in a manner that could have been improved upon, even within
the context of the open field system. It appears then that those
farmers who were inefficient managers were the ones that remained in the open field villages, and that, by implication, those
who were efficient managers were the ones that enclosed. This is,
of course, another of the dumb peasant arguments : the ones in
the open fields were less intelligent than the rest, even if all of
them were reasonably prudent and intelligent. And if the arguments of modern writers are more sophisticated, it is clear that
the underlying logic is still that of the rather slow-minded farmer
plodding around his scattered strips.

It would therefore seem that the traditional hypotheses are
difficult to uphold on several grounds. Even if we do accept one
of the traditional explanations for the origin of the scattering, we
have to reject them as theories of the preservation of scattered
strips, for they all imply that reasonably intelligent farmers should
have exchanged land with each other to avoid the costs of scat-

[39] *Ibid.*, p. 209.

tering. There is no accounting for the fact that the open field system survived until the eighteenth century – it really ought to have disappeared quite early.[40]

The mirror image of this is that the traditional hypotheses do not account for the 'unscattering' of the strips in a satisfactory way either. For example, the new techniques of the Agricultural Revolution were introduced in the open field farms as well;[41] hence, it follows that enclosure was not at all necessary for new techniques. In other words, scattering is quite compatible with the introduction of alternative methods of production, and therefore it is not possible to argue that consolidation of scattered holdings is a *sine qua non* for the attainment of efficient agriculture. The wool trade explanation fares better in this respect, for when pasturage was vastly increased, at the expense of the arable, the scattered strips would gradually be very much reduced in importance. It is quite likely that, after enclosure for sheep-raising, just enough arable was left to feed a greatly diminished population on farms that were very small compared with the village itself; and that scattering, for whatever reasons it had been imposed, was abolished as a natural consequence.

Nor can the traditional hypotheses give an adequate explanation for the extensive communal regulations or for the fact that property was owned communally by the members of the open field village. Against the argument that communal ownership is a hereditary feature of our culture it is again necessary to ask why this element persisted so doggedly in the face of the obvious costs of communal ownership. Why was the waste not divided among the members of the village at one time or another? In the traditional version, one allegedly inefficient element is made to account for the next inefficiency. Joan Thirsk writes :

. . . as the parcels of each cultivator became more and more scattered, regulations had to be introduced to ensure that all had access

---

[40] Curtler states this honestly : 'The disadvantages of the open-field system were numerous, and the only wonder is that it lasted as long as it did.' *Enclosure and Redistribution*, pp. 112–13.

[41] See e.g., W. G. Hoskins, *Provincial England*, London, 1963, p. 153; M. A. Havinden, *Agricultural Progress in Open-field Oxfordshire*, reprinted in W. E. Minchinton (ed.), *Essays in Agrarian History*, Newton Abbot, Devon, 1968, pp. 156–9; E. Kerridge, *The Agricultural Revolution*, London, 1967, p. 19; Yelling, *Common Field and Enclosure, passim.*

to their own land and to water, and that meadows and ploughland were protected from damage by stock. The community was drawn together by sheer necessity to cooperate in the control of farming practices. All the fields were brought together into two or three large units. A regular crop rotation was agreed by all and it became possible to organize more efficiently the grazing of stubble and aftermath.[42]

In this argument the inefficiencies inherent in the scattering are made to account for further inefficiencies. One is compelled to ask why the village members did not exchange their plots with one another, for, with the elimination of scattering, communal organization of cropping and harvesting, along with communal grazing, could be abolished, and as a consequence, the random breeding and contamination by disease would cease also.

Furthermore, again beyond the wool trade explanation of Tudor enclosures, there is nothing in the traditional hypotheses that can give a clue as to the correct timing of enclosure. Since it was not really at all necessary to resort to enclosure to achieve the benefits of the new techniques, there is nothing in the explanation for the changeover to a new and better system that can account for why a particular village chose to undertake the change at a particular time. Various *ad hoc* idiosyncrasies in the village environment can be brought in – such as a particularly foresighted lord who saw the benefits of the new techniques, or progressive farmers insisting on ridding themselves of the yoke of their more backward neighbors – but there are no *general* conditions that can be pointed to in order to determine why enclosure occurred at a particular date rather than at some other.

As a comprehensive solution to the problem of understanding the vagaries of open fields and enclosure it seems that the 'traditional hypotheses' are inadequate. This brief analysis can only serve to show the complexity of the task involved, especially with respect to the data available to substantiate any claim to having a better explanation than the ones discarded here. As yet, there is simply no accepted 'general' outline of all the features of the open field system or the enclosure movement that historians can settle on as the received doctrine. Perhaps such a state of affairs will

---

[42] Joan Thirsk, 'The common fields', p. 9.

never be achieved. In the meantime, we can only search for better explanations for the stylized facts that we do have.

The Marxian exploitation hypotheses. Although perhaps few writers on open fields and enclosures can be called pure Marxian, in the sense that their explanations run exclusively, or even mainly, in terms such as class struggle, exploitation, surplus value and so on, there is often a strong element of Marxian thought discernible in the background of many discussions on the decline of the open field system. For example, many authors stress the importance of the landed gentry, the main promoters of enclosures, in Parliament, and their role in legislating enclosure bills aimed at the expulsion of small farmers; others stress the importance of the close links between the rise of industries, with their need for more labor, and the enclosures in the eighteenth century; still others point to the great differences in living conditions between the landed nobility and gentry, on the one hand, and the tenants working the land, on the other, in terms which have strong overtones of Marxian indignation. It appears important, therefore, to attempt to distill the outlines of the 'pure' Marxian exploitation model in a fashion that will allow for an examination of its inherent logic, and of its empirical relevance to the open field system and the enclosure movement.

While attempting to do so, it must be remembered that there is no such pure Marxian exploitation model presented by any one writer in a manner that can be readily discussed. Our first task must therefore be to piece together, as best as we might, the broad outlines of an exploitation version of Marx's ideas. Having done so, we must then proceed to discuss the difficulties of a logical and empirical nature embodied in the writings of Marx. The bare outlines of the Marxian exploitation hypotheses are very simple : before enclosure the lords exploited the peasants by exorbitant fines, and enclosure was a means of intensifying that exploitation. This was made possible by the fact that the lords constituted the dominant stratum, politically and economically, in the hierarchy, and by sheer force and coercion lived off the peasants. The robbery of the peasants that constituted the enclosure movement was feasible because the landowning classes dominated **Parliament**.

Such are the outlines of the Marxian exploitation hypotheses. To reiterate, the story itself is very simple, and the message clear. However, it remains to be seen whether the logic of the hypothesis, in as rough a presentation of it as this, is consistent with the open field system and the enclosure movement, and whether the historical facts bear out the continuing story of exploitation of peasants. The first issue involves the definition of exploitation, and the relationship between surplus value and rent, and the second the question of whether class is a concept applicable to medieval England, the role of custom, and the role of Parliament in the enclosure movement.

To understand exactly what is meant by exploitation in Marx it is necessary to understand the concept of surplus value. Schumpeter gives a lucid account :

Marx's exploitation theory may be put like this. Labor (the 'labor power' of the workman, not his services) is in a capitalist society a commodity. Therefore, its value is equal to the number of labor hours which are embodied in it. How many labor hours are embodied in the laborer? Well, the 'socially necessary' amount of labour hours it takes to rear, train, feed, house him, and so on. Suppose that this labor quantity, referred to the labor days of his active span of life, figures out at four hours per day. But the 'capitalist' who bought his 'labor power' – Marx did not go quite so far as to say that the 'capitalist' buys laborers as he could buy shares, though this is the implication – makes him work six hours a day. Four of these six being enough to replace the value of all the goods that went to the laborer, or the variable capital advanced to him (v), the two additional hours produced Surplus Value (s), the *Mehrwert*. For these two hours the 'capitalist' has not given any compensation. They constitute 'unpaid labor'. To the extent that the laborer works hours that are unpaid in this sense, he is exploited at the rate s/v.[43]

However, this is the relationship with respect to the 'capitalist', and it remains to be seen whether this concept is also applicable to the situation in agriculture. This involves the concept of rent. Marx recognized two different concepts of rent, ground rent and differential rent. On ground rent he says :

In practice everything appears naturally as ground-rent that is paid for in the form of lease money by the tenant to the landlord for the

[43] J. A. Schumpeter, *History of Economic Analysis*, Oxford, 1954, p. 649.

permission of cultivating the soil. Whatever may be the composition of this tribute, whatever may be its sources, it has this in common with real ground-rent that the monopoly of the so-called owner of a piece of the globe enables him to levy such a tribute and impose a tax. This tribute furthermore shares with the real ground-rent the fact that it determines the price of the land, which, as we have indicated above, is nothing but the capitalized income from the lease of the land. . . . All ground-rent is surplus-value, the product of surplus-labor. In its undeveloped form, as natural rent (rent in kind), it is as yet directly the surplus-product itself.[44]

This defines ground rent. With differential rent Marx understands the same thing as does Ricardo, that is, that the price of land may be affected by the natural conditions of location and fertility, which may affect the return to the owner of the land. In a sense, therefore, differential rent is also a form of exploitation. Marx says :

Value is labor. Therefore surplus-value cannot be land. The absolute fertility of the soil accomplishes no more than that a certain quantity of labor produces a certain product conditioned upon the natural fertility of the soil. The difference in the fertility of the soil brings it about that the same quantities of labor and capital, hence, the same value, express themselves in different quantities of agricultural products, so that these products have different individual values. The equalization of these individual values into market-values is responsible for the fact that the 'advantages of fertile soil over inferior soil . . . are transferred from the cultivator or consumer to the landlord.' (Ricardo, *Principles*, p. 6.)[45]

Since the land produces nothing in itself without labor, the resulting output must be attributed to labor alone. Therefore, even the differential rent, if kept by the owner of the land, is surplus value, since it is clearly attributable to labor but appropriated by the owner of the land. It is surplus value, for it represents output over and above what is required to keep the laborer alive, and hence it is part of the exploitation of the fruits of labor.

The first point to note about exploitation through the appropriation of surplus value is that there are certain limits to the rate of exploitation achievable through this method. Some writers maintain that the peasant would be exploited until he is living at

[44] K. Marx, *Capital*, vol. III, pp. 732, 743.
[45] *Ibid.*, p. 948.

subsistence level. However, Marx himself did not drive the exploitation concept that far :

Some historians have expressed astonishment that it should be possible for the forced labourers, or serfs, to acquire any independent property, or relatively speaking, wealth, under such circumstances, since the direct producer is not an owner, but only a possessor, and since all his surplus labor belongs legally to the landlord. However, it is evident that tradition must play a very powerful role in the primitive and undeveloped circumstances, upon which this relation in social production and the corresponding mode of production are based. . . . Society assumes this form by the repeated reproduction of the same mode of production, where the process of production stagnates and with it the corresponding social relations. If this continues for some time, this order fortifies itself by custom and tradition and is finally sanctioned as an expressed law. Since the form of this surplus labor, of forced labor, rests upon the imperfect development of all productive powers of society, and upon the crudeness of the methods of labor itself, it will naturally absorb a much smaller portion, relatively, of the total labor of the direct producers than under developed modes of production, particularly under the capitalist mode of production.[46]

Since labor is not bought and sold by contractual arrangements under a mode of production which is not capitalism, pure contractual forms will not determine the wage rate. Instead, custom and tradition will play an important role in the intimate and personal relationship between a lord and his serfs, and this will hamper the rate of exploitation that the lord can subject his tenants to. In this way we can determine the upper bound, as it were, to the rate of exploitation possible in the open field system. Is there also a lower bound? Schumpeter points out that there ought to be competition for the surplus value obtainable from labor :

According to Marx, surplus value is a costless gain, like Ricardian rent. It might be thought that such a gain would induce individual capitalists – whose individual contributions to the total output of their industries is too small to influence prices – to expand output until the surplus falls to zero. This conclusion is indeed inescapable so long as we keep to the scheme of a stationary process; such a process could not be in equilibrium until the surplus is eliminated. But we

[46] *Ibid.*, pp. 921–2.

may save the situation by taking account of the fact that Marx thought primarily of an evolutionary process in which the surplus, though it has a tendency to vanish at any given time, is being incessantly recreated. Or else we might drop the assumption of perfect competition, though the surplus we may salvage in this way will be quite different from Marx's.[47]

That is to say, unless the landlord continuously devises new means of exploiting his tenants, there will be a danger that all the surplus value he can capture from the peasants will be competed away by neighboring lords who offer better terms on their land. This gives perspective to the horror tales commonly told of how the ruling classes subjected the peasantry to new and seemingly harsher payments all the time. These can be interpreted as a defensive adaptation by which the landlords guaranteed that they achieved the rent which allowed them to continue to live as idlers, and does not necessarily imply that the rate of exploitation also increased throughout the whole period we are considering here. This appears as a more attractive solution than the alternative one, of assuming that the lords banded together into cartels or monopolies of exploiters, although this is an argument which is sometimes resorted to in order to explain the rate of exploitation of the peasantry. In this manner, the exploitation hypothesis provides a concept which can serve as the *primus motor* of historical evolution. In order to keep to the rate of exploitation necessary to preserve them as a class, the rulers are forced by competition to change the mode of production currently in use, and this inevitably forces changes on society. In a way, this also provides a rationale for the specific *timing* of change : old ways of exploitation have run dry, and new ones must be invented. Perhaps it can also be said that the theory points to the precise sources of change, for if surplus value is reduced by competition, the ground rent will be competed to zero. However, the differential rent cannot be competed away, being incorporated in the nature of the soil. Unless the differential rent is capitalized into the value of the estate, one can therefore infer that the lands yielding no differential rent on the margin will be the first to force change on society by devising new methods for exploitation as the ground rent approaches zero.

[47] Schumpeter, *History of Economic Analysis*, p. 651.

In a sense, this version of the exploitation hypothesis comes very close to a model along classical economic lines, with profit maximizing landlords eagerly on the look-out for new prospects of increasing their earnings, yet all of them ultimately gaining nothing except what is a normal rate of return to capital. The major difference between the two models then lies in the treatment of capital: in Marx *all* of the product above the socially necessary minimum is surplus value, even if the amount thus extracted is kept at bay by custom or Schumpeterian competition; in the classical model all payments to capital are determined by considerations of marginal productivity. Naturally, this is a fundamental difference in outlook between the two frameworks of thought, but one that should not totally distract attention from the very important similarities between them. There seems to be no support in Marx for the harsh exploitation version – the one that sees the peasantry as brutally mistreated and kept at starvation level. There is still exploitation, of course, but of a significantly different magnitude from what the harsher version would imply. Thus, it still follows that class struggle over surplus value is the fundamental contradiction that initiates change. On logical grounds, therefore, the exploitation hypothesis of the evolution of the open field system cannot be rejected. The problems involved seem to be rather of a more normative kind, pertaining to the desirability of private versus collective ownership of capital and the results this has for the construction of the model. Conceptually, one must make a decision as to whether surplus value is exploitation or payment to capital – except for this distinction there appears to be little fundamental difference between the Marxian exploitation hypothesis and a classical perfect competition model with productive capital.

It follows perhaps that the decision whether to accept or reject the Marxian exploitation hypothesis as an account of the evolution of the open field system and the enclosure movement will have to be made with reference to the empirical evidence: can the hypothesis account for more of the stylized facts, as it were, than alternative explanations? Two considerations seem to be germane here. The first is whether the basic concept of class, in the Marxian sense, can be applied to medieval agriculture, and whether the broad outlines of a class struggle can thereby be discerned. The second concerns the question of predictability of actual events from the logic of the model, that is, whether the model can discriminate between different evolutionary scenarios or is consistent with any conceivable event.

It is extremely doubtful that the concept of class has much to contribute to the understanding of relations in agriculture in medieval England. On class, Dobb says :

The common interest which constitutes a certain social grouping a class, in the sense of which we have been speaking, does not derive from a quantitative similarity of income, as is sometimes supposed : a class does not necessarily consist of people on the same income level, nor are people at, or near, a given income level necessarily united by identity of aims. Nor is it sufficient to say simply that a class consists of those who derive their income from a common source; although it is source rather than size of income that is here important. In this context one must be referring to something quite fundamental concerning the roots which a social group has in a particular society : namely to the relationship in which the group as a whole stands to the process of production and hence to other sections of society.[48]

Our problem is whether we can regard the lords of the manors in a way that would allow us to label them a class in this sense : that they have a common interest in preserving their relationship to the rest of society as regards the process of production with respect to the exploitation of surplus value. Here there are fundamentally three different relationships. The first concerns the internal structure of the manor or the open field village itself : can we regard this basic social unit as a class society in any relevant sense ?

In the open field village there were several status groups : nobility, gentlemen, yeomen, husbandmen, tradesmen, laborers, and paupers. But the clear delineations that would be necessary to identify them as classes simply did not exist. One status group shades over into the next; and, besides, their identification depends more on occupation than on social status, which is of course subject to one additional qualification : most tradesmen, yeomen, and husbandmen dabbled in several occupations at once, so that even status groups may be difficult to discern clearly.

[48] M. Dobb, *Studies in the Development of Capitalism*, London, 1946, p. 15.

Servants and laborers were not employed simply by the nobility and the gentry, but by less substantial farmers as well:[49] the point at which the ruling class ends and the oppressed begins would be impossible to determine. Not only were some of them relatively wealthy, but they also had an interest in preserving the present mode of production as they lived off surplus value from hired labor. One cannot say that the lord alone constituted the ruling class in the village community in the sense that he alone stood to gain from exploiting surplus value.

Nor did the lord have that ultimate power necessary to usurp the peasantry at will.[50] We have seen above how Marx stressed the importance of custom and tradition serving as an impediment to wanton impositions of dues on the peasantry. It would be wrong to look at the lord simply as someone devoted to the idea of squeezing the last drop of blood from his tenants' starving families. What Bloch says about France is equally valid for England:

Nothing could be more misleading than to dwell exclusively on the economic aspects of the relationship between a lord and his men, however important they may seem. For the lord was not merely a director of an undertaking; he was also a leader. He had power of command over his tenants, levied his armies from them as occasion demanded, and in return gave them his protection. . . . Many a Frankish king or French baron if asked what his land brought him would have answered like the Highlander who said 'five hundred men.'[51]

In such circumstances it would be unlikely that the lord would even attempt to wring the life-blood out of his tenants with any great conviction. It is important not to forget that the relationships between the members of the village community must have been very personal, as the villages were usually very small. Laslett estimates a 'typical' village, if there be such a thing, to contain

---

[49] See P. Laslett, *The World We Have Lost*, New York, 2nd ed., 1971, p. 67.

[50] Pollock and Maitland, *The History of English Law*, vol. I, pp. 361–2.

[51] M. Bloch, 'The Rise of Dependent Cultivation and Seignorial Institutions', in Postan (ed.), *The Cambridge Economic History of Europe*, 2nd ed., vol. I, p. 72. See also R. H. Hilton, *A Medieval Society: The West Midlands at the End of the Thirteenth Century*, London, 1966, pp. 24–5, and R. H. Tawney, *The Agrarian Problem in the Sixteenth Century*, London, 1912, *passim*.

about 200 persons of all categories. In such a small community, with people living their entire lives side by side, any excess of abuse would be difficult to uphold for sheer humanitarian reasons. Yet there were other obstacles. The lord could not simply dictate custom and tradition, especially in a country such as England where common law is the foundation of the legal system. Titow states: 'Thirteenth century villeinage was neither arbitrary nor unpredictable; even though it was deprived of the protection of the royal courts, the lord knew what he was entitled to and the peasants knew what to expect, since such matters were governed by the Custom of the Manor which was binding on landlords and peasants alike.'[52]

This must especially have been the case since all the tenants owed suit of court to the lord, and the jury was composed of village members. To believe that the lord manipulated custom and the courts at will is to negate an important body of historical facts. The tools necessary for unlimited exploitation of the peasants were not in the hands of the lords.

Secondly, it would be wrong to infer that the lords of manors, dispersed all over England, constituted a class with community of interest. The lords fought each other over property and peasants. It was almost impossible for a lord to prevent a tenant from leaving, and even more difficult to recapture him with the help of other lords. We have stressed the importance of competition between lords as part of the explanation of how ground rent is diminished, thus forcing new methods of exploitation to come to the fore. This can only be interpreted as a recognition of the fact that the lords never acted as a class with community of interests.[53]

Thirdly, it would not be correct to infer that the lords acted in union with the capitalists in the rising manufacturing industries when they evicted the peasants from the land so as to create a

---

[52] J. Z. Titow, *English Rural Society; 1200–1350*, London, 1969, p. 58.

[53] Lord Ernle, *English Farming*, p. 354, stresses this point as well: 'We feel naturally inclined to think and to speak of the village community in opposition to the lord and to notice all points which shows its self-dependent character. But in practice the institution would hardly have lived such a long life and played such a prominent part if it had acted only or even chiefly as a bulwark against the feudal owner. Its development has to be accounted for to a great extent by the fact that lord and village had many interests in common. They were natural allies in regard to the higher manorial officers.'

reserve army of labor after the enclosure movement.[54] One of the main contentions of the Hammonds, in their book *The Village Labourer*, was that the landed classes through their domination of Parliament expropriated the land of the poorer tenants. Later research has discredited this view, and the current theme in the literature is rather to wonder at the fact that complaints about the fairness of the decisions of the commissioners effecting enclosures were not much more numerous. Clearly, one finds here little support for the contention that the landlords were able to use Parliament as an instrument in their class struggle against the laboring classes.

There is a further difficulty with the exploitation interpretation of the course of historical events. Exploitation is the confiscation of the fruits of unpaid labor, which through competition between the exploiting class members is forced to approach zero. But new methods of exploitation are constantly being introduced, and this precipitates social change. These are the clues provided to construct a theory of social change that can account both for the element of timing and for the direction of the change itself. That is to say, we should be able to point to the features of society at any one time that make ground rent approach zero, and should therefore also be able to predict more or less precisely when change ought to come about. If the landlords find that peasants are moving from their manor to others with lower rents, but that lowering the rents will make it impossible for the landlord to sustain his standard of living off the lower amount of surplus labor thus exploited, he may decide to enclose. A change in property rights structure may allow for a change in the rents, so that a new form of exploitation replaces the old. Similarly, by looking at the different alternatives open to the lord at that time, we ought to be able to predict that he will choose enclosure – as opposed, for example, to the imposition of some new technique or crop regulations or trading arrangements – as the change which will best serve in preserving the maximum rate of exploitation.

Here the exploitation hypothesis has little to contribute. It

---

[54] See J. D. Chambers, 'Enclosure and labour supply in the Industrial Revolution', *The Economic History Review*, 2nd series, vol. V, 1952–3, *passim*, for an analysis that lays this ghost to rest.

would seem to argue simply *post hoc, ergo propter hoc,* without providing any specific insights into the elements of timing or the direction of change. Enclosure occurs because the lords must find the best alternative method of exploitation when their current expropriation of surplus labor is falling off. Yet how can we ascertain that the forces committing the lords to this course of action ought not to have occurred some centuries or decades earlier? Until there is a detailed account of the factors that determined the course of the fluctuations of ground rent, we are at a loss in our efforts to determine anything about the timing of the change itself. We have a general condition, but nothing that shows its applicability in the particular context. It is the same with the course of change : how can we really be certain that *enclosure,* and not some alternative novel mode of expropriation, was actually the preferred course of action? It is because of its inability to provide answers to such specific questions that the exploitation hypothesis must be abandoned as an explanation of the open field system and the enclosure movement. If the historical evolution takes one course, the explanation will be that this is due to the ruling classes' search for new methods of exploitation of the laboring classes; indeed, this will be the explanation for any changes that occurred in the past.

The exploitation hypothesis is of little or no help in accounting for the stylized facts of the open field system or for their disappearance. It gives us no clues at all, for instance, to the scattering of the strips in the open field system; and, by implication, it offers no clues for consolidation either. Neither does it point the way to an acceptable explanation of the peculiar mixture of private and collective property rights in the open field village, and therefore does not account for the privatization of ownership and decisionmaking after enclosure.

In conclusion, it would seem that the exploitation hypothesis leaves too many loose strands for provision of a coherent story, and we move on to some other theories of open fields and enclosures.

Some alternative explanations. In the face of all the difficulties encountered by the Marxian exploitation idea, it is perhaps not surprising that an alternative Marxian model, purportedly truer to the spirit of Marx himself, has been advanced.[55] This alternative model attempts to avoid the difficulties inherent in the exploitation concept. Instead, it focuses on the fact that our historical knowledge of social relationships in medieval society does not allow us to infer that peasants were universally downtrodden by cruel and selfish landlords. The basic tenets of the model are that, before enclosure, landlords were not profit maximizers. Rather, they put emphasis on the number of men they controlled; and hold the view that with the advent of a commercial spirit, landlords became more concerned with profit making; that this led them to appropriate the 'agricultural surplus' from the peasants necessary for the new-found objective of 'primitive accumulation'; and that this, in turn, was the prerequisite for the growth of the capitalist society that superseded the feudal relationships of reciprocity between strata in the hierarchy. The means by which this appropriation was achieved was enclosure, and it necessitated an efficient utilization of resources. So after enclosure, excess labor was dismissed from the agricultural sector.[56] Although enclosure may have meant an increase in total income for the community, the laboring classes were made worse off by enclosure as the landlords claimed a greater proportion of total income than before. Enclosure, therefore, was a two-edged sword. It meant, on the one hand, that total production increased as existing resources were utilized more efficiently, but, on the other hand that the share of income going to the laboring classes went down and made them worse off in absolute terms.

Although the basic idea of this model is very simple, the actual presentation of the analytical framework, within which the propositions are derived, contains some crucial oversimplifications that make a detailed criticism of the argument unnecessary. It will be sufficient to mention in passing some of the major diffi-

[55] J. Cohen and M. L. Weitzman, 'A Marxian Model of Enclosures', pp. 287–336.

[56] Cohen and Weitzman maintain that the guiding principle of resource allocation and income distribution before enclosure was that wages were set equal to the average product, not to the marginal product. As a result, labor was inefficiently allocated on the given quantity of land, and total output was paid out in wages, except for an arbitrary but small head tax to the lord.

culties with the approach insofar as they concern the explanation of the structure of the open field villages.

The model has nothing to say on the origin and persistence of scattering – indeed, scattering is not of any relevance to the model at all. Cohen and Weitzman have therefore no explanation to offer for the 'unscattering' of the strips in the enclosure movement either. Nor can the model explain the mixture of private and collective rights that were so characteristic of the open field village. The model explicitly assumes that the open field village was a non-exclusive resource, in Cheung's terminology,[57] i.e., that newcomers could not be excluded. This cannot give an adequate account for the existence of private rights of tillage in the strips of the open fields, nor indeed for the existence of a formal collective decisionmaking body in the village court. The very extensive rules set up in the open field economy, and the great homogeneity of those rules between villages in different parts of the country, show that they were set up to cope with some specific resource allocation problem. It does medieval man an injustice to say, as do Cohen and Weitzman, that there was no rationality and intelligence involved in the making of those rules.

Furthermore, the model's explanation of enclosure is deficient at best. The claim is that there was a change in behavior on the part of the lords. It is difficult to explain why this occurred at all, since the change to a 'commercial spirit' is outside the purview of the model. Consequently, the theory is at a loss to explain the timing of enclosure. It appears inexplicable that enclosure should take several hundreds of years to accomplish. If a rearrangement of legal institutions suddenly allowed one lord to maximize his profits, why did all other lords not do the same? Why was the agricultural surplus not claimed immediately when conditions allowed, so that the primitive accumulation could get under way? Cohen and Weitzman have no explanation for this. Indeed, one is left with the question why there is not one single benefactor of the peasantry left in the whole of England today. Surely some lord, at least, would have cared enough for his men to keep the system in its old form? It is difficult to find a better label for this

---

[57] S. N. S. Cheung, 'The structure of a contract and the theory of a non-exclusive resource', *Journal of Law and Economics*, vol. XIII, April 1970, pp. 49–70.

c

model than the 'dumb lord' model, to complement our 'dumb peasant' model above.

The foregoing explanations of the existence of the open field system and its disappearance in the enclosure movement have all relied on factors other than those of market transactions or economic incentives to account for the course of historical events. Instead, they have attempted to account for the system in terms of social interrelations, historical circumstances, inertia on the part of the lords or the peasants, inadequate institutions, greed, and ignorance. In contemporary economic theory, such concepts play little part. One will turn in vain to modern textbooks for an exposition of behavior determined by the factors just mentioned. Instead, economic models rely on a basic assumption of utility, wealth, or profit maximization under known constraints. It is perhaps little wonder, then, that the 'New Economic History' marks not only the advent of quantitative methods into a field where literary style has always counted for more than the ability to handle algebra, but also the surge of formally structured models based on postulates of maximizing behavior. The next model discussed here is an example in kind.[58]

The work referred to is that of McCloskey. His basic idea is that an acceptable explanation for the scattering of the strips requires us to assume that the farmers who tilled them had powerful *economic* reasons for adhering to this peculiar arrangement. The purpose of setting up a formal model of the open field system, therefore, must be to identify those advantages of scattering that can be shown to outweigh its costs. A convincing case must be made for scattering as a predictable market response. The circumstances under which an ordinary English peasant had to carry out his daily tasks were so uncertain, the argument goes, that it would be natural for him to pay a substantial insurance premium in order to decrease the unpredictable fluctuations in his income. That is to say, a postulate of risk aversion appears to provide the one explanation that can account for the persistence of scattering, since in that case the mechanism of market ex-

[58] McCloskey's works in this area are : 'English open fields as behavior towards risk', in Uselding, *Research in Economic History*, 'The persistence of English common fields', and 'The economics of enclosure', both in Parker and Jones, *European Peasants*.

change would no longer serve to erode, but would preserve, discontinuous holdings. This would be the case as long as the peasant could be reasonably certain that a random disturbance to the yield on one part of his holdings would not equally affect the other parts scattered in different locations; in other words, as long as the correlation between outputs on different plots was low. McCloskey maintains that conditions were so variable in the typical English village that this was the case.

Furthermore, alternative methods of insurance simply did not exist. A peasant interested in reducing the variability of his income would have to accept the inefficiencies inherent in scattering as the only way to achieve this goal. Any other way to satisfy a desire for risk aversion would prove more costly, and scattering is therefore consistent with lively trading in the market for land. Two necessary conditions for accepting scattering as a viable method of avoiding risk are thereby established: scattering is technically effective in reducing risk, and it is economically efficient, being the cheapest method available under the circumstances.

The next step is to develop an explicit model of how risk aversion can yield scattering as a function of variables that can be empirically estimated. McCloskey employs a model developed in the finance literature for the choice of the composition of a portfolio of assets. The method is the so-called mean-variance approach where it is assumed that the utility functions of the peasants considering scattering are expressible in terms of the mean of the prospective income and its variance alone. He thereafter proceeds to estimate the degree of scattering that this approach would predict, given the estimates that one can obtain of output variance, correlation coefficients between crops on different plots, elasticity of the yield per acre with respect to the size of the strip, and other quantifiable and empirically determinable magnitudes. The number that the theoretical method predicts can then be compared with the actual number of plots the typical farmer had in the open field. McCloskey claims that he has been very successful in his attempts to establish the credibility of his approach: his empirical estimates predict almost exactly the number of plots he would want.

McCloskey's research is still in progress, and detailed criticism is therefore premature. There is still a long way to go to establish the credibility of the approach, for finding a model that predicts the correct number of strips in each field for the typical peasant really only amounts to a very small step. To convince us that risk aversion really was at the heart of scattering, several additional elements must be accounted for. For example, McCloskey has not shown, other than through general assertions, that a farmer who was risk averse, and thus scattered his plots, actually in the long run would do better than one who did not. This is, of course, a crucial point. Simple logic would tell us that unless, over a period of years with variations in income, a farmer who scatters actually attains a net income higher than one who does not scatter, market forces and competition, on which McCloskey lays such great stress, would soon eliminate the economic foundation for the scatterer, and only those who did not scatter would then survive. That is, it would seem incumbent on McCloskey to show that those who did not scatter would not be able to sustain the farming venture, presumably through falling below a minimum income in years of harvest failure. So far McCloskey has not addressed this question, except in a vague manner so different from his precise numerical estimates of various parameters relevant to the model.

An additional element that McCloskey simply ignores is the intricate technological relationship between the production of grain and the production of livestock that formed such an important element in a typical open field village. He addresses only the production of grain in the arable, but ignores the fact that many farmers actually did attain or, if they did not, easily could have attained a sizeable portion of their income from the production of livestock. Unless it can be shown that there is a positive or zero correlation between the prices of livestock and those of various grains, McCloskey has not successfully discarded the possibility that the production of both of these two interrelated outputs actually could have reduced the variability in yearly income streams.

This is an argument about the availability of alternative ways of insuring against too low incomes, but it is by no means the only one that McCloskey ignores. There are at least two other counterfactuals that deserve serious consideration. The first is that, in spite of the ridicule that McCloskey heaps on the lack of what he refers to as 'charity',[59] many landlords were forced, willy-nilly, to be charitable to their tenants in years of low yields. It is a well-known fact that, when tenants were in arrears, many landlords were forced to refrain from collecting a rent. Now, if you wish to carry an argument about the efficiency of market organization to its logical conclusion in this context, it is easy to demonstrate that, if rents were determined competitively (as McCloskey would have it), profit maximizing landlords would simply overcharge their tenants a little in good or average years, and undercharge them or not charge at all in bad times. By this method they would still receive a competitively determined average rent over several years on all their property. If there was little correlation between the variability in the incomes of individual farmers, such a method would even be quite compatible with little or no variation in income for the landlord himself. The point about this counter-factual is that, McCloskey's unsubstantiated assertions notwith-standing, there did indeed exist a viable market for insurance, through the lord–tenant relationship : the lord could, and, if his-torians are right, did act as a veritable lending institution in bad years.

The other counterfactual that McCloskey ignores is one that relies on the demonstration by Reid that it is possible to attain any degree of risk dispersion desirable through a combination of labor and land contracts.[60] It is a well known fact that pure rent contracts lay the whole risk of accepting income variations on the tenant, for independent of actual crop results he must pay the contracted yearly rent. On the other hand, a labor contract puts the risk on the shoulders of the landlord, for he contracts for the labor, and has to pay out of an uncertain crop outcome. He therefore bears the total risk. What Reid has shown, in the con-text of share tenancy as an alternative way of risk sharing, is that it is perfectly feasible for the landlord and the tenant to get any risk sharing they care to achieve by a combination of labor and land contracts. Is this an argument that has relevance in the context of the open fields?

[59] McCloskey, 'English open fields', p. 64.

[60] J. Reid, 'Share-cropping as an understandable market response : The post-bellum South', *Journal of Economic History*, vol. XXXIII, 1973, pp. 106–30.

It is, of course, easy to argue that it should have been perfectly possible for the lord and the tenants to undertake such cross-contracting in labor and land. It would be difficult to uphold an argument that says that this would not have been feasible, in view of the fact that landlords and farmers could and did hire labor, and lords and farmers alike rented land, or easily could have done so. Since such contracts constituted the fundamental relationship between the lord and the tenants anyway, it is difficult to see why it would not have been possible for them to attain risk dispersion, had they so desired, in this simple extension of already existing contractual agreements. However, depending on how one views the institution of serfdom, there is yet another twist to this argument, which at present can only be stated as an hypothesis to be further explored. It is a fact of history that serfdom was not a one-way street: both parties, the lord and his serf, accepted mutual obligations when the oath of fealty was given and accepted. The serf became the lord's man, and was bound to him for the rest of his life. But what that version of the story so often ignores is that the lord was also bound to his man by a tie that could not be undone at will. The implication here is simply that a lord who could not collect rent from a villein in a bad year could not just evict him – he had responsibility towards his tenant and villein.

In addition, however, the villein did work for the lord, and received therefore a payment, which can be viewed as consisting of two parts: one in kind payment made by the lord to his men on days of boonwork, and a monetary one in terms of the adjustment of the rent on the land that the lord could, if he wished, extend to his tenants. The point about this is that serfdom indeed was an institution that not only had the possibility of establishing, but most probably *did* establish, the kind of land–labor contractual combinations that yielded risk sharing between lord and peasant. If this view is correct, then McCloskey must not only explain why he contends that no alternative ways of risk sharing were available to open field farmers, but he must face the awkward fact that historical evolution had discarded an institution that could, in principle, yield perfect risk sharing in favor of one that could do so only imperfectly, at a net cost in terms of output which was higher than that yielded by the discarded institution.

There is yet another element that McCloskey has not so far accounted for adequately: enclosure, and how it is consistent with risk aversion as a force yielding scattering. For after enclosure scattering disappears. Unless McCloskey wishes to argue that enclosure was due to a change in the attitude towards risk, he must then contend that some alternative mechanism arose that allowed for the desire for risk aversion to be satisfied. McCloskey has not yet shown that such a mechanism did exist.

In addition, McCloskey treats scattering as an independent element of the open field system. We have argued above that this is too limited a view: scattering can only be understood in the context of the open field system as a totality of interrelated elements. Thus McCloskey has no explanation for the interrelatedness of scattering and the other stylized facts that were listed above. He claims: 'The analysis, in any case, confirms the hypothesis that scattering was the root cause of whatever loss occurred: scattering, with its attendant inefficiencies implied common grazing, with more inefficiencies, and scattering and common grazing together implied communal cropping, with still more.'[61]

Quite apart from the fact that no econometric analysis ever 'confirms' any hypothesis, this statement is not fully taken into account by McCloskey. When he calculates the loss of income that the average farmer sustains because of scattering, he does not take into account the 'inefficiencies' of common grazing, nor those of communal cropping. That is to say, when he reports a figure of three percent as the loss in annual income due to scattering, he is understating the true loss of income by an unknown amount. He himself argues that it was large, but does not reason through the implications for the optimal amount of scattering. Until this is done, there is no acceptable account for the other stylized facts, nor is there any discussion of the internal logic of a system so rich in institutional inefficiencies. However, McCloskey's final analysis is still awaited.

These criticisms of McCloskey's theory of scattering should not detract from the very valuable contributions to our understanding of the open field system that McCloskey has achieved. His analysis of various arguments for why scattering persisted is unsurpassed in the literature. Perhaps his most important influence is that he, ultimately, would seem responsible for the attention the open field system has received in the last few years from economists. Indirectly, McCloskey gave the impetus for this study, as well as for several papers by other authors in various journals. For that we shall owe him a debt of gratitude even if he should be unsuccessful in establishing the credibility of the risk sharing approach to scattering. It remains the only work into the optimal degree of scattering, and as such it is an important contribution. The defects of the theory only pertain to the claims of causation that McCloskey advances, and even if those claims are exaggerated, the work itself will stand up.

3

PROPERTY RIGHTS, TRANSACTION COSTS, AND INSTITUTIONS

Having completed the destructive task of tearing down what others have built up the time has come to begin the construction of an alternative framework, free of the detectable defects of existing structures which purport to be models of the open field system. In this chapter, we shall lay the conceptual and methodological foundations for a theory that will be shown to yield a consistent explanation of the riddles left unsolved from the previous chapter. However, since the theoretical framework to be used is based on a new and developing branch of economic theory, one that has not yet crystallized into a set of final propositions and theorems, it will be necessary to extend the existing analysis in order to demonstrate how a property rights and transaction costs approach can further the understanding of the functioning of institutions. In this context, the concept of transaction costs, its specific relationship to economic institutions, and what is meant by 'choosing' institutional arrangements will form these extensions.

First, however, a proper justification must be advanced for choosing the particular methodology that will be employed here, for it is not immediately clear why we shall have to turn to such a relatively undeveloped branch of economic theory as the property rights paradigm in order to find the keys to the open field system. Returning to the stylized facts of the representative open field village in the previous chapter, it will be seen that these facts can be condensed into four main determinants. First, there is the nature of the resource endowments. This includes such features as the quality of the soil, topography of the location of the village, climate, availability of fertilizers, access to waterways, and other such features of the natural environment that are beyond the control of man. Second, there is the technology of production. In technical economic terms this simply means the relationship between the productive inputs and the outputs produced. It may be conceived of as the available knowledge of crop rotations and breeding programs, irrigation methods and the effects of alternative fertilizers, and of ways to control erosion, etc. Third, there are the ownership rights. On one hand this involves the question of wealth distribution, or the entitlement to the income from the assets available in the village. But, on the other hand, it is also a question of the rights to various activities : who has access to the commons, who must pay a fine for what infraction, who has the right to transfer or consolidate, and on what conditions, and related questions. Fourth, we come to the institutional structure. This concerns the formation and function of the court or the village meeting, its scope of decisionmaking and control, and the method by which it reached collective decisions. In order to account for the open field system as a system, we must show how all these pieces interlock in a way that produces a cohesive and intelligible pattern.

To accomplish this, we must first make a classification of the variables into exogenous and endogenous, to use the technical language of the economist. That is to say, in order to explain the open field system as a construct of its decisionmakers, we must determine which of the variables that together constitute the salient features of the system actually were within the control of the members of the open field village. These will then be the variables that we shall analyze to see if we can find an explanation of the choice of solutions to the particular problems faced by the open field village.

Examining the fourfold classification above, we note that both the nature of the resource endowments and the technology of production were elements that were essentially beyond the control of any individual village. The quality of the soil and the climate and such variables were given to the village, and formed part of the conditions that determined the layout of the village and the way it was tilled. This is true also for technology of production : the state of the art is inherited from the past, and learnt from adjoining villages, in the way that knowledge of how to transform raw materials into finished form is always given to us. On the other hand, the remaining two classes of variables, the ownership rights and the institutional arrangements were elements within the powers of decision of the village members. Since there is nothing, for example, in the communal ownership of the commons or the setting up of a court that is either determined by divine powers or imposed on the village by a higher governmental authority, we must conclude that these were matters that the village itself decided on, and chose to implement and preserve for some definite reason. Furthermore, enclosure and the establishment of private, individual rights were always viable alternatives to the preservation of the open field system. The endogenous variables, to be determined by the analysis, therefore, are the ownership rights and the institutional arrangements.

This is the reason for choosing a property rights approach in this essay. For property rights are considered endogenous variables of an economic system, and we are beginning to attain a more formal understanding of the determinants of the choice of property rights systems. Also, the link between property rights and institutions is a very intimate one; indeed, they may be considered as part and parcel of the same phenomenological complex. In the emerging property rights literature we therefore have a conceptual framework to draw on that immediately deals with problems similar to the ones at hand. It is to be noted how little there is in standard or traditional economics that deals with these issues : the dominant paradigm of modern economics, Walrasian general equilibrium systems, is completely void of institutional structures, for reasons that will perhaps become obvious as our analysis proceeds. Moreover, apart from the property rights method, there is virtually nothing in economics that can justify one institutional construct as superior to another.

In standard economic theory of the determination of prices and quantities in a market context, modern received theory proceeds in the following manner. Take as given the quantity and quality of productive resource endowments, the behavior of individual agents (utility, wealth, or profit maximization with a known objective function), and technology of production. With technology and endowments as exogenous constraints, quantities demanded and supplied, and their prices, are determined with simple optimization procedures : maximize utility or wealth or profits, subject to the known constraints. Except for the statement that ownership of the initial endowments is well known and uncontroversially distributed among people, nothing is said about the institutional content of the conceptual framework. All other institutions, such as firms, governments, corporations, etc., are submerged and/or taken as given exogenously in standard general equilibrium analysis. What the property rights approach can be said to accomplish is to insert an intermediate step into this method, one that justifies the presence of economic institutions and their treatment as endogenous variables. Again, taking endowments and technology of production as given quantities, the property rights approach introduces one additional exogenous element : in contradistinction to traditional choice theory, the analysis explicitly recognizes the existence of transaction costs. For the moment, we may simply interpret this concept to mean real resource costs of moving from the initial distribution of endowments to the final allocation of resources for production and consumption.

In principle, for every given set of initial endowments, tastes, and technology, there will always exist a multitude of final allocations in consumption and production. What the notion of transaction costs introduces is a way of comparing such final allocations : in terms of real resources used up in the process of exchanging goods and services, one final allocation may be superior if less resources are expended in moving to that rather than to any other final allocation. However, transaction costs are related to the institutional environment : choosing one institutional environment may, therefore, give you more or less transaction costs than some other institutional environment. Hence the object of choice is to find that institutional environment that maximizes the value of the final allocation by minimizing the transaction costs associated with moving from the initial to the final distribution of resources.

This is the intermediate step inserted by the property rights approach. It may be said that standard constrained choice theory starts when the problem just referred to has already been solved. For standard theory takes as given, and as only implicitly understood, a set of institutions that govern and regulate the process of exchange, and then proceeds to show that, given these institutions and the other variables, one final allocation is superior to all others. Viewed in this light, property rights theory constitutes an addition to received doctrine, and not an alien or foreign element. Consequently, many of the standard concepts of economic theory remain useful, while the vista opens up for new questions and, we hope, for new answers. In Chapter 6 we shall address the question of whether some accepted results of standard theory will have to be amended when property rights are considered endogenous and transaction costs are explicitly recognized.

This is the conceptual background for the founding of a theory of the choice of economic institutions. Within the logic of this framework, some initial results have been attained, and we shall have to scrutinize these, insofar as they have a bearing on the problem of the choice of institutions and ownership rights in the open field system. In addition, we shall have to analyze the transaction cost concept in some detail, in order to show how various methods for dealing with them work. The next few sections of this chapter will be concerned with these two issues.

The meaning of the notion of property rights

In a study of comparative economic systems, for example, it is common to refer to the differences between various economic systems, notably capitalistic and socialistic ones, in terms of the ownership of resources. This is the perhaps traditional interpretation of what is to be understood by the term 'property right'. In order to elucidate how the phrase has come, in modern economic interpretation, to have a much wider scope than this, it is necessary to inquire into the fundamental meaning of the concept of 'ownership of resources'.

On one hand, ownership implies possession : the thing that is owned is in some way physically controlled by the owner. Some think that this is the fundamental and only significant difference between different economic systems, and will inquire into the ethical or efficiency characteristics of state versus private ownership, for example. Such an approach clouds the important fact that physical possession is only a limited part of what must be understood about ownership, for a thing is not possessed in and for itself, but only to be used to satisfy certain desires on the part of the owner. On the other hand, therefore, we must include in the notion of ownership the uses to which the thing, or the economic asset, to use the technical phrase, can actually be put. Physical possession is only part of the story. The more important element turns out to be the rights of undertaking various actions with respect to the use of the asset. In their survey of the property rights literature, Furubotn and Pejovich therefore define the notion of property rights in the following manner :

. . . property rights do not refer to relations between man and things, but, rather, to the sanctioned behavioral relations among men that arise from the existence of things and pertain to their use. Property rights assignments specify the norms of behavior with respect to things that each and every person must observe in his interactions with other persons, or bear the cost for nonobservance. The prevailing system of property rights in the community can be described, then, as the set of economic and social relations defining the position of each individual with respect to the utilization of scarce resources.[1]

Property rights, then, really refer to the rights of decisionmaking with respect to an asset that are recognized within any given social setting. Property rights constitute a collection of social codes, legal remedies, and societal protection or sanctions. It follows therefore that the crucial distinction between capitalism and socialism is not so much state versus private possession, but the fact that, in a capitalistic system, a specific individual is allowed to make decisions that the government makes in a socialistic system. Related to this is the important notion of attenuation : all rights of usership or decisionmaking, whether state or private, are always subject to certain limitations.[2] There is no such thing as absolute ownership, not even in an economic system characterized by complete private ownership. For even a system with private ownership of scarce resources will prohibit certain actions in the interest of the community as a whole. For example, the right to use an implement to damage somebody else physically has always been circumscribed in every known society. Sometimes, even in the most capitalistic societies, the right to damage others

[1] E. G. Furubotn and S. Pejovich, 'Property rights and economic theory : A survey of recent literature', *Journal of Economic Literature*, vol. X, no. 4, 1972, pp. 1137–62. The quotation is from p. 1139.

[2] *Ibid.*, p. 1140, for a further discussion of this important concept.

economically is also circumscribed by law. Thus, an owner who holds land in fee simple may still be prohibited from erecting certain structures on his land, or from putting the land to uses that have obvious side effects on his neighbors. What is to be understood by property rights is therefore not simply physical possession of an object, but the uses to which an economic asset can be put; and this is a question of what social sanctions exist.[3] With property rights we should, therefore, really understand attenuated decisionmaking rights.

One further observation is essential: what gives relevance to the study of different property rights systems is the important fact that human behavior is predictable. The implication is that if we impose various attenuated property rights systems, or various restrictions on decisionmaking rights with respect to the use of economic assets, the result is predictable and consistent behavior. Naturally, this is a prerequisite for a purposeful study of property rights: if people did not behave rationally, i.e., predictably and consistently, property rights would not be a meaningful phenomenon to study.

This is the ultimate justification for the study of property rights systems, for if it can be shown that humans respond in different and predictable ways to decisionmaking restrictions imposed on them, in one way or another, then we might be able to construe a case for comparing the relative merits of various property rights systems. It might be shown, for example, that if our goal is to maximize the value of output, given inputs and technology, then a certain mixture of property right is preferable to any other mix. Questions such as this have led to the study of problems that have come to be known under the heading 'law and economics', whereby economists and lawyers alike attempt to find out if some laws actually perform better than others. However, this branch of property rights theory is not immediately relevant in the present context, and will not be explored.

But another application of this methodology has great relevance for any theory that claims to portray the open field system.

---

[3] For example, it is common to understand by the notion of private property the right to use an asset as the owner sees fit, to exclude other users from laying claim to the asset, and to exchange the rights in the asset with anyone the owner sees fit. Usually, all these rights are attenuated in one way or the other in every known economic system.

Specifically, there exists in economics a firmly entrenched contention that, given the self-interested acquisitiveness of human behavior, collective usership rights in a scarce economic asset are inevitably associated with inefficient resource allocation. One of the many areas where this seems to be of practical importance is aquaculture; since no one has an exclusive right to most fishing areas, there is the danger that over-entry of fishermen results in overfishing, to the point where the extinction of certain species sometimes becomes a real possibility. There are many statements of this in the economics literature, but the best and most consistent is perhaps the one by Cheung.[4] By amending Cournot's oligopoly model, Cheung is able to show that, with successive entry of users to a non-exclusive resource, usership will extend to the point where rent is completely dissipated, i.e. to the point where the fish become depleted. Naturally, this is inconsistent with the efficient use of a scarce resource, and in his simple context Cheung is able to show that private ownership of the fishing grounds would maximize the rent rather than dissipate it as in the case of non-exclusivity. Thus, collective ownership and non-exclusivity are associated with inefficient resource allocation.

In the property rights literature, this argument has been extended to provide a justification for private ownership of scarce resources. A well known paper by Demsetz will serve to illustrate this as well as the use of the transactions cost concept in the analysis of a choice of property rights problem.[5] Demsetz examines a case where changes in market conditions induce the establishment of private property rights in a productive resource previously subject only to collective rights. The reference is to a Canadian Indian tribe, and it is sufficiently interesting for our purposes that we shall summarize and, to some extent, extend the argument.

Before the advent of the fur trade to the Labrador Peninsula, there were only non-exclusive or collective rights in the land. The hunting of animals, which was carried out mainly for food, was open to everybody without restrictions. This did not create any

---

[4] S. N. S. Cheung, 'The structure of a contract and the theory of a non-exclusive resource', *Journal of Law and Economics*, vol. XIII, April 1970, pp. 49–70.

[5] H. Demsetz, 'Towards a theory of property rights', *American Economic Review*, vol. LVII, 1967, pp. 347–59.

noticeable problems as the fur animals were not scarce. However, with the sharp rise of the fur trade which followed the arrival of the French, circumstances changed radically, for the animals now became a source of income from the sale of the fur. By the beginning of the eighteenth century, property rights had become established in different tracts of land, and individuals were allocated the rights of hunting within certain tracts. This constituted a change from collective ownership of a productive resource, the land on which the animals fed, into private ownership of that same resource. Why did private property rights become established?

Demsetz shows that with collective ownership, there is an externality present. When an individual catches a fur animal, he receives the whole economic benefit of the hunt, but the costs are borne by all other hunters in that there now are fewer animals left for them to catch. Thus collective ownership provides the incentive for the individual to over-use the collectively owned resource, for he gains the whole benefit but bears only part of the cost. By the same token, there will be an incentive present under collective ownership to invest too little in the replacement of the animals; for if a hunter decides not to capture a young animal or a mother with cubs, he will bear the full cost of not getting the fur now, but the benefits of getting more fur in the future are bestowed on all hunters since everybody has the right to hunt. Therefore, with collective ownership, there will be an externality present as between the hunters. One hunter's actions will impose costs or benefits on the other hunters that will not enter ino his calculations when he decides how much he will hunt. The consequence is an incentive to overconsume and underinvest in the productive resource.

This is what happened to the Indians when the fur trade became intensified. The scale of hunting increased and the animals were no longer used simply to provide for self-sufficiency. Clearly, as the fur animals became relatively more scarce, i.e., as their market value increased, the resulting incentives to hunt more may have increased the significance of the externality so as to make the overconsumption and underinvestment economically important. Demsetz shows that in this situation, private property rights represent a cheaper way of attaining the correct allocation of resources than do collective property rights. For with private ownership rights, there will be no incentive for overhunting. Thus the theory predicts that private property rights should replace collective rights when the resource under consideration becomes sufficiently scarce.

This would seem to yield an extremely simple criterion for the choice of property rights systems : all we need is to examine the relative scarcity of existing economic resources. If they be scarce, then let us establish private rights. If they be abundant, we may preserve free access and collective ownership. However, this simplicity is deceptive, and we shall have in due course several occasions to question the simple criterion Demsetz puts forward. Still, Desmetz' analysis is at present the only consistent argument for the choice of property rights systems, and has to be taken as a starting point, if nothing else. It may be noted, however, how closely related his analysis is to the very harsh indictment of the commons as a source of inefficiency in the open field system so often found in the historical literature.[6] The assertion is that the communal access to the grazing areas led to overgrazing, random breeding of animals, the spread of diseases, and generally poor upkeep. The frequently used argument that enclosure would have improved efficiency is really nothing but an application of the Cheung–Demsetz principle that private ownership of a scarce source is superior to collective.

The relationship between Cheung's and Demsetz' arguments is therefore very close, and the preceding discussion will have made it clear that what prevents an efficient solution in both cases is the inability on the part of the economic agents involved to contract voluntarily for a restriction of the use that is excessive. It would be in the interest of all parties to stipulate appropriate bounds on the activities of all claimants to the scarce resource. This would prevent the rent from dissipating, and would ensure proper reinvestment. What prevents this is the right of anybody to enter into and use the resource for himself. The self-interested acquisitive behavior of humans will result in inefficiency becoming unavoidable under collective ownership.

[6] G. Hardin, 'The tragedy of the commons', *Science*, vol. 162, 1968, pp. 1243–8.

This analysis revolves around the crucial notion of transaction costs. In the example above from Demsetz, a conceptually feasible alternative to private property would be for the bands to retain communal ownership, but agree to limit overhunting. The case would be the same in the fisheries example of Cheung in which the fishermen could abide by an agreement not to overfish. However, such a voluntarily agreed stipulation would entail policing costs for the interested parties : they would have to make sure that all hunters or fishermen complied with the agreement. That is to say, an exchange whereby each economic agent restricts his activity in the common interest in exchange for a guaranteed access to a plentiful resource in the future is not costless : it must be policed, as every single hunter or fisherman has an incentive to cheat on the agreement and appropriate some economic gain for himself at the expense of the others. Similarly, they would have to agree to keep outsiders out, and this is an additional cost of implementing the agreement of limiting overhunting or overfishing. It is really the presence of such transaction costs that prevent an efficient allocation of resources from being established.

On the other hand, there is a crucial distinction between fur hunting and aquaculture. We have seen that the Labrador Indians resolved the difficulties associated with collective ownership by establishing what amounted to private rights in the land, but that solution is usually not feasible in fishing waters, for reasons that anyone who has heard about the plight of whales is familiar with. Because of the migratory habits of fish, it would be necessary to patrol and police vast expanses of water in order to protect and preserve private ownership rights in certain fish or fishing grounds. In aquaculture, private ownership rights may often prove infeasible, whereas such rights may be very efficient in other kinds of assets. The reason is that transaction costs, in this case policing costs of ownership rights, may not be independent of the nature of the resource itself or of the production technology that determines the utilization of the resource. If these transaction costs are of a certain nature, then private ownership rights may prove an efficient alternative to collective rights; but there are cases in which private rights are not a practical alternative. Private rights are not a panacea.

The conclusion that private ownership rights in a scarce asset are unambiguously superior to collective rights is, therefore, not always justified. The analysis of the relative efficiency of the two forms of ownership and control will have to include a discussion of the following elements : the relative scarcity of the asset, the costs of establishing and making effective any agreement between the users of the asset to avoid over-use, and the costs of establishing and protecting ownership rights in the asset. The methodological implication is that, in order to understand the open field system, we must make plausible the exact nature of these and other so-called transaction costs that faced the decisionmakers of the open field village if we are to understand why they chose one particular system over another.

Property rights and institutions

There is one additional issue that can conveniently be dealt with before we proceed with a more detailed discussion of transaction costs. This is the relationship between economic institutions, viewed as entities separable from the property rights concept, and transaction costs. We shall return to a discussion of the precise relationship between property rights and economic institutions and attempt to justify the distinction between them in Chapter 6. For the moment, it will be sufficient to characterize property rights as attenuated decisionmaking rights, in the manner of the preceding discussion, and to think of economic institutions as actual economic decisionmakers. The distinction between human agents and economic institutions may then be thought of as that between primary and secondary agents : humans exist and act by virtue of being born into the system – but institutions, the secondary agents, are man-made creations. For the moment, we shall confine our attention to one such institution : the classical firm. The firm is an agent that makes decisions with respect to the allocation of scarce assets, notably the capital belonging to its owners, and it is a secondary agent, as it is a construct of human ingenuity.

Why do such secondary agents arise? Consider the question posed by Coase, in a paper that has become a classic in the theory of the firm : if you do believe, as economists perhaps ought to, that the price mechanism serves to allocate resources efficiently,

why do we not see one worker on a production line buy the un-
finished product from the previous worker on the line, add his
work to it, and then sell it on to the next worker on the line?
Why is it that workers in a firm obey commands issued by an
authority instead of relating to each other through market exchange? The answer Coase supplies us is :

It can, I think, be assumed that the distinguishing mark of the firm is
the supersession of the price mechanism. . . . Our task is to attempt
to discover why a firm emerges at all in a specialized exchange
economy. . . . The main reason why it is profitable to establish a firm
would seem to be that there is a cost of using the price mechanism.[7]

Again we are back to transaction costs. If there were no costs
of transaction, then the logic of our well established price theory
would indeed make us predict that the organization of production
ought to take place through the price mechanism, and not be
'superseded' by authoritarian decisionmaking. However, if there
are costs of using the price mechanism, then those costs can be
saved if we can find a way that controls activity directly, through
orders, instead of through bidding and bargaining. This may then
give us an account for the existence of 'conscious power' outside
the market mechanism, and still allow us to retain the logic of
prices as signals for resource allocation.

So far, however, nothing has been said about what the costs of
using the price mechanism might be. Coase points to several such
costs : the cost of assembling and disseminating information about
prices, the cost of negotiating a contract for each separate transaction, and the cost of renegotiating short-term contracts when
the contract term is up.[8] All these costs can be eliminated or at
least reduced by an entrepreneur. In this way, economic theory
can provide a justification for the emergence of a firm that supersedes the price mechanism; and we should look for the nature
of the transaction costs the firm is able to save on.

However, in spite of the logically pleasing nature of the argument that the concept of transaction costs yields a rationale for

[7] R. H. Coase, 'The nature of the firm', originally in *Economica*, vol. IV,
1937, pp. 386–405; reprinted in G. J. Stigler and K. E. Boulding (eds.),
Readings in Price Theory, American Economic Association, 1953, pp.
331–51. Quotation taken from pp. 334, 335, 336.

[8] Coase, 'The nature of the firm', pp. 336–7.

the existence of the firm as an organization replacing the market mechanism, it must be admitted that Coase's formulation is rather vague. There is little specificity in his treatment of the precise way in which the firm saves on transaction costs. Fortunately, there exists a more recent treatment of this particular problem : the contribution by Alchian and Demsetz. Although the conceptualization of the firm as reducing the costs of production by avoiding certain market transactions is clearly borrowed from Coase, Alchian and Demsetz introduce some elements that make their version of the theory different in some substantial respects. Primarily, they question the view of the firm as an 'island of conscious power' that rules by authority.[9] Whereas Coase views the firm as 'superseding the price mechanism', Alchian and Demsetz show that, from a conceptual viewpoint, the firm indeed constitutes much the same contractual arrangement as the ordinary contracts that we see being established across any market. A consumer is empowered by the market mechanism to fire his butcher in precisely the same manner as the firm is empowered by the contract it has entered into with its employee to fire that employee. Thus, the firm does not supersede or replace the market : rather, it serves as a perfect example of the functioning of the market mechanism.

Since the distinguishing feature of a firm, therefore, is that it is the result of a sequence of contractual agreements between economic agents, we shall have to look at the establishment of economic organizations, of which the firm is but an example, as the result of economic transactions. That is to say, organizations are endogenous, man-made creations, on the one hand, and they constitute a collection of decisionmaking rights, or property rights, on the other. What describes a firm are the decisions that it is empowered to make, and we may simply look at a firm as a particular collection of certain attenuated property rights. Indeed, we may generalize this view to include all economic organizations, including those of the open field system. It will be noted how in the description of our representative village the institutional structure was concerned with scopes of decisionmaking both for the members of the village and the court that guided their affairs.

---

[9] A. A. Alchian and H. Demsetz, 'Production, information costs, and economic organization', *American Economic Review*, vol. LXII, 1972, p. 777.

Again, the methodological implication is that we shall have to look for an economic justification for the court and its functions in the nature of the particular transaction costs that were relevant in the open field villages. Also, we should look at the organizations within the villages as a result of contractual agreements between the village members. We should describe the function of the court in terms of the attenuated decisionmaking powers, i.e., by defining the property rights vested in the court. We must also recognize clearly that what makes the court able to acquire such property rights or decisionmaking powers is the fact that the members of the village, explicitly or implicitly, in their obedience of court rulings, are themselves the ultimate source of the viability of the organization they set up and preserve.

The interpretation of transaction costs

The crucial point to emerge from the preceding analysis is clearly that the pivotal element around which an understanding of the choice of property rights systems and the emergence of economic institutions must turn is the concept of transaction costs. In order to understand, therefore, how the open field system was designed to cope with certain transaction costs emerging in its particular environment, we must attempt to gain a deeper understanding of the precise nature of the concept. In view of the crucial role transaction costs play in the context of institutions, it is rather remarkable that no systematic, general analysis exists of the nature of such costs. In recent years, the concept has come to achieve a rather prominent place. It has become a catch-all phrase for unspecified interferences with the price mechanism, but it has also been shown that an understanding of this concept is necessary for the foundations of monetary theory.[10]

There is a useful line of thought on transaction costs in Coase's definition :

In order to carry out a market transaction it is necessary to discover who it is that one wishes to deal with, to inform people that one wishes to deal and on what terms, to conduct negotiations leading up to a bargain, to draw up the contract, to undertake the inspection

[10] This was brought to the attention of modern eyes by Clower, 'Foundations of monetary theory', reprinted in R. W. Clower (ed.), *Monetary Theory*, Harmondsworth, 1969, and his introduction to the same.

needed to make sure that the terms of the contract are being observed, and so on.[11]

Since it would appear that this definition is difficult to handle with mathematical tools, it is consequently not surprising that this notion is prevalent among thoroughly non-mathematical writers – notably those who treat issues in law and economics. However, to show its usefulness in the context of an analysis of economic institutions, it is necesary to take this definition of transaction costs a little further.

A natural classification of transaction costs consistent with Coase's definition can be obtained from the different phases of the exchange process itself. In order for an exchange between two parties to be set up, it is necessary that the two search each other out, which is costly in terms of time and resources. If the search is successful and the parties make contact, they must inform each other of the exchange opportunity that may be present, and the conveying of such information will again require resources. If there are several economic agents on either side of the potential bargain to be struck, some costs of decisionmaking will be incurred before the terms of trade can be decided on. Often such agreeable terms can only be determined after costly bargaining between the parties involved. After the trade has been decided on, there will be the costs of policing and monitoring that the other party carries out his obligations as determined by the terms of the contract, and of enforcing the agreement reached. These, then, represent the first approximation to a workable concept of transaction costs : search and information costs, bargaining and decision costs, policing and enforcement costs.

Yet this functional taxonomy of different transaction costs is unnecessarily elaborate. Fundamentally, the three classes reduce to a single one – for they all represent resource losses due to lack of information. Both search and information costs owe their existence to imperfect information about the existence and location of trading opportunities or about the quality or other characteristics of items available for trade. The case is the same for bargaining and decision costs, which represent resources spent in

[11] R. H. Coase, 'The problem of social cost', *Journal of Law and Economics*, vol. III, 1960, p. 15.

finding out the desire of economic agents to participate in trading at certain prices and conditions. What is being revealed in a bargaining situation is information about willingness to trade on certain conditions, and decision costs are resources spent in determining whether the terms of the trade are mutually agreeable. Policing and enforcement costs are incurred because there is lack of knowledge as to whether one (or both) of the parties involved in the agreement will violate his part of the bargain : if there were adequate foreknowledge on this part, these costs could be avoided by contractual stipulations or by declining to trade with agents who would be known to avoid fulfilling their obligations. Therefore, it is really necessary to talk only about one type of transaction cost : resource losses incurred due to imperfect information.

This gives a rather general definition of what is to be understood by transaction costs. We now understand it to mean real resource losses occurring because of imperfect knowledge on the part of some trader or economic agent. With appropriate knowledge, about prices or locations or qualities of goods as well as other traders, some actions might have been undertaken that in the absence of that knowledge were not, and as a result, a less than optimal allocation of scarce resources results. The efficiency loss due to actions undertaken with wrong or imperfect information will then be referred to as the loss due to the presence of transaction costs, for if information were to be had at zero cost (i.e., if transaction costs were zero) then nothing could prevent resources from going to their highest value in use. With transaction costs, we must then understand set-up and transfer costs, information costs about prices, qualities and desired transactions, including costs of detecting and avoiding cheating and strategic bargaining behavior.

We have yet to address the question of how exactly it is that property rights and institutions can contribute to the minimization of such transaction costs. To take the example of property rights first, it is easy to show how this reduces the cost of making an exchange. Suppose ownership rights in an asset were not clearly defined, and that, to use the wider interpretation of what is meant by property rights, the decisionmaking rights with respect to that asset were not clearly specified either. It follows that if an economic agent contemplated purchase and acquisition of that asset, he would be faced with two uncertainty problems : given that there are no precise ownership rights in the asset defined for a given individual, he would not know exactly who would lay claim to the asset. There would be uncertainty as to who has rights in the asset and therefore a prospective buyer would not be able to make a precise enough calculation of the value of the usership to himself of the asset. The case is precisely the same if the established rules about usership of the asset are unclear. This would be a situation in which the prospective buyer faces uncertainty with respect to the social rules and regulations that attenuate the ownership, once acquired. Again, in the face of such uncertainty, it would be difficult to make a precise present value calculation of the benefit stream that the asset will yield over its economic life time.

Naturally, both these elements of uncertainty could be reduced with the acquisition of better information about the probabilities of additional claimants demanding the asset and the likelihood of social rules changing, respectively. The crucial consideration, however, is that although such information can be had, it would require resources to collect it and ascertain its reliability. This is what transaction costs amount to : in the face of such uncertainty, additional resources will have to be spent by economic agents to ascertain unknown probabilities of certain outcomes. By defining private exclusive ownership rights, and by making certain that the social codes defining acceptable uses of the asset are known and enforced, the economic system as a whole can save on costs of individual information gathering. At least, this will be the case if the private ownership rights and the codes of social behavior can be defined and protected at a cost that is less than what would be incurred by economic agents in their absence. If a trader knows, with certainty, that a seller of an asset has private exclusive rights in an item for sale, then he knows that he will be able to exclude other claimants to that item, and so he will be able to make a better calculation of the value of the asset. The case is the same with respect to usership rights. Knowing the 'rules of the game' established by social decisionmaking institutions will alleviate the problems of calculating correctly the present value of the benefit stream to an asset. Since in this way fewer resources have to be spent on gaining information, it is easily

understood how private ownership rights and well defined usership rules increase the value of economic goods available for trade,
and so lead to a different and improved allocation of resources.
So there is a good case for clearly defined property rights contributing to the minimization of transaction costs. Note, however,
that the logic of this argument does not necessarily imply that
private ownership should always be established in a scarce
economic asset, for there are costs of establishing and protecting
ownership rights and of policing compliance with socially determined attenuations of decisionmaking rights. It is consequently
possible to envisage a situation in which these costs may be higher
than the transaction costs saved. In such a situation, it may be
preferable, in the name of efficient allocation of resources, that
no property rights be defined at all.

Similarly, it may be shown how various institutions contribute
to the minimization, if not complete elimination, of certain transaction costs. If we wish, we may look at transaction costs as the
economic rationale for certain political institutions. Congress or
parliament, in a system of representative democracy, represents
a way for a great number of individuals to bargain with each
other, and thereby constitutes a forum for decisionmaking. The
fact that, in congressional or parliamentary voting procedures, a
specific rule is imposed is simply a way of reducing the uncertainties involved. Without such a rule, the bargaining power of a
strong senator or member might distort the outcome even more
than is presently presumed to be the case. The State also sets up
a court system that sometimes serves as a bargaining agent for
parties who disagree on the interpretation of various rules or
commitments that they have entered into, and thus contributes to
the saving on costs of agreement between these parties. Furthermore, the court system and the police power of the State serve to
reduce the costs of enforcing the social codes of the polity, compared to what would be the case if private contracting had to be
relied on.

State institutions of this kind are not, by far, the only examples
which show how the economic system attempts to cope with
various transaction costs. Many non-governmental institutions
also exist because they serve to decrease the costs for traders to
meet, agree, and comply with established rules. A few examples

will suffice. Banks may be said to specialize in searching out likely investment opportunities for savers, to inform them of available investment alternatives, to bargain the lending and borrowing rates between savers and investors, and to police that borrowers actually fulfill their obligations. Real estate agents serve both to locate and inform buyers about available property and to bargain an agreement between seller and buyer. Lawyers serve as bargaining agents for their clients, both in court and often outside the courtroom, in order to keep the involved parties out of court. They also specialize in ensuring that agreements are actually adhered to, and so reduce the costs for the involved parties of policing compliance themselves. Insofar as money is legal tender, it may be regarded as a means of reducing a number of transaction costs. In the absence of money, i.e., in a situation in which economic agents would have to rely on barter, it is well known that there is a problem of mutual coincidence of wants. Money does not require such coincidence, but represents a means of indirect exchange. In such a manner, it saves on costs of search, for the probability of finding someone offering money in exchange for a certain economic good is higher than finding someone offering a specific good, as the case would have to be under barter. Money also contributes to the minimization of bargaining and decision costs, for there is less scope for disagreement on the terms of trade. Since only one good and money are involved in monetary exchange, as opposed to at least two goods in barter, bargaining will be confined to fewer goods than under barter.

Such simple examples could be multiplied, but the point has already been made clear. The interpretation of the transaction costs notion as costs due to uncertainty can account for a variety of economic phenomena, and is therefore a potentially very useful tool of analysis. An additional clarification must be made, however. The preceding classification of transaction costs derives its structure from the sequence of transactions that constitutes the heart of a mutually beneficial exchange : search and information costs, bargaining and decision costs, policing and implementation costs. This is satisfactory as far as it goes, but it leaves one class of transaction costs yet to be defined. This concerns the notion of property rights itself, for such rights do not emerge out of thin air. Rather, they have to be established within the economic system, as part of the complex of relationships that define the preconditions for the interactions between the economic agents of the system.

In the process of defining property rights, the economic system must make two interrelated decisions – both of which have already been referred to in an oblique manner. The first is to decide on the distribution of wealth : who shall have the rights to owner-ship of the scarce economic resources even before, as it were, trading and contracting can begin. The second refers to the allo-cative function of property rights : they confer incentives on the decisionmakers within the economic system, for the attenuated rights determine what can be done and what cannot be done with any specific economic asset. It is clear, therefore, that we must deal with costs of making the 'transactions' that constitute the defining of a social contract that sets the preconditions for the ensuing economic trading game. We can separate them into two parts : one set of decisions must be treated as endogenous for the system, and constitute the exogenous conditions for each trading agent in the resulting set of trades; the second set of decisions is made in the context of the making of these trades. What we have considered in the foregoing is only the nature of the transaction costs pertaining to this second set of decisions.

In a way, this is inevitable, for little can be said at the present stage of economic and political theory about the first set of deci-sions. These decisions constitute the making of a constitution for not only an economic system with its trading agents, but more fundamentally also for a political system with the implicit ethical and moral judgments that must be made in that context. How-ever, in spite of the rather intimidating breadth of the problems involved, a few things can be said about the transaction costs involved in the first set of decisions.

For property rights to be determined, it is not sufficient for a political body to enunciate a decree by which Mr. A becomes the owner of asset Z, and so on, but also to find a way in which to make effective and practical any such decision. That is to say, the political body making a decision about wealth distribution must also ensure that its decisions be made effective : property rights must be protected, for otherwise their definition is of little consequence as they become ineffective, both as a mechanism for the redistribution of wealth and as a tool for devising an incentive system. The political decision of defining ownership rights must be policed in an effective manner. That is, a mechanism for ensuring that rights actually become exclusive must be presumed to exist already in order for a discussion of the efficiency characteristics of various property rights assignments to be meaningful.

In the next chapter, we shall proceed to show how the notion of transaction costs developed in the preceding has obvious relevance for the understanding of the property rights and institutions characteristic of the open field system. It will also become evident how the distinction between primary and secondary agents is of importance for a proper understanding of the establishment of the decisionmaking scope and powers of the village meeting or the manorial court. Furthermore, the division of the decisions about property rights into two conceptual stages will also have immediate relevance for an analysis of the open field system, for many legally established rules and regulations were beyond the influence of the typical open field village, representing a constitution, as it were, given to it by higher governmental authority. Thus the tools surveyed and developed in this chapter will have immediate application for an understanding of the intricacies of the open field system.

The question of efficient institutions

Through transaction costs, we may in the preceding manner justify a suggestion that property rights and economic institutions serve to improve economic efficiency. We may push this inquiry one step further. Is there such a thing as an 'efficient' set of institutions? Will the economic system guarantee the establishment of such efficient institutions, or will there sometimes be 'inefficient' economic institutions prevailing in the economy and if so, how often? These are complex questions, and at this stage of the analysis no answer will be attempted. Instead, we shall content ourselves with some observations on that part of the property rights literature that has a bearing on this issue. Since we have already exemplified the tendency towards the establishment of efficient institutions, of which the firm is such a lucid illustration, we shall limit the discussion to the question of whether the economic system sometimes establishes inappropriate or inefficient organizations and/or property rights.

Examples of this kind may be easily found in the literature on regulation of industries. It is recognized in modern treatises that much regulation exists, not so much for the protection of the public, as for the benefit of the regulated industry itself. By sheltering itself under a 'regulatory umbrella', the industry may be able to shield itself from competition, and thereby in effect establish a monopoly that receives the protection and sanction of government agencies and legal institutions. This is the so-called capture theory of economic regulation.

Another example of a very similar process is the adjustment by decisionmakers in an industry where the profit rate is regulated by the government. Alchian and Kessel have shown that in such a situation, utility maximization on the part of the managing officers of the regulated firm will take the place of profit maximization.[12] They will increase the costs of the firm by hiring attractive but sometimes incompetent secretaries, by creating perquisites for the management, by building luxurious offices, and by other means, all designed to capture part of the rent that is dissipated through profit controls. However, such activities are inefficient, as a normal competitive process, with profit maximization, would lead to their elimination.

Another ingenious example can be found in the modern theory of money and its implications for the optimal structure of the banking system.[13] It has been shown that an institutional framework consistent with the complete elimination of unanticipated inflation, and hence also of the negative consequences normally associated with inflation, would be one where each individual bank, as in the beginning of the history of the banking industry, is allowed to issue its own money, backed by some real commodity such as gold. The implication is that the government monopoly on money production that is such a prevalent feature of almost all modern societies is really an example of an inefficient and undesirable institution.

[12] A. A. Alchian and R. Kessel, 'Competition, monopoly, and the pursuit of pecuniary gain', *Aspects of Labor Economics*, Princeton, 1962.

[13] B. Klein, 'Competitive interest payment on bank deposits', *American Economic Review*, vol. 64.

In an application of property rights theory to history, North and Thomas also point to similar processes in attempting to find an explanation of why certain countries underwent an industrial revolution before others.[14] Initially, they point out that growth is a question of efficient institutions. Those countries that had efficient institutions grew, those that did not stagnated. One important reason for these differences lay in the way the State used its taxing power to finance current expenditure.[15] A government would often grant monopoly rights to some individual in exchange for a loan or other favors. It is clear that such privileges for certain individuals were inefficient from the standpoint of society as a whole, for they limited the degree of competition in an artificial manner. Implicit in this explanation of the creation of inefficient institutions is an argument that relies on transaction costs. The statement that State-granted privileges for a few were inefficient for society is equivalent to stating that it would be feasible, in principle, for those who lose from the existence of the monopoly to bribe those who gain from it. Thus, everyone could theoretically gain from the abolishing of the monopoly rights, including the ruler who granted them in the first place. The ultimate reason that such bribery does not occur is the existence of costs of transacting in getting the beneficiaries together, in making them reveal their true evaluations of the benefits from the elimination of the monopoly, and in making them pay their promised contributions to the collective bribe. It may therefore be cheaper for one beneficiary, the State or the ruler, to create inefficient institutions, although, from the viewpoint of the system as a whole, this is an undesirable course of action.

Why, then, does the economic system tolerate the establishment and perpetuation of such – and other – inefficient economic institutions? The answer would seem to be that the institutions in question are not inefficient from the viewpoint of those who personally benefit from them, such as the industries benefiting from regulation or the government financing some taxation through the monopoly on money production. The point is that although, for the system as a whole, these institutions are ineffi-

[14] D. C. North and R. P. Thomas, *The Rise of the Western World*, Cambridge, 1973.

[15] North and Thomas, *The Rise of the Western World*, pp. 98–9.

cient, they are not inefficient in redistributing income to certain groups or individuals who directly or indirectly benefit from them. The book by North and Thomas has many examples of institutions that were efficient in this particular respect – in enabling certain groups in the system to use the structure to achieve an improvement in their own wealth position. Indeed, the basic argument of North and Thomas as to why certain countries did not enter into a process of sustained growth is that such self-interested parties that succeeded in creating inefficient institutions for their own benefit thwarted the potential that existed for growth in several countries. Thus it would have been in the interest of everyone else to change the institutional structure so that growth could occur. This did not happen, so the inefficient institutions remained, and benefited the few at the expense of the many.

It might be thought that it is possible to give a similar interpretation of the open field system, that it was an inefficient institution that served the purpose of creating income for the lord of the manor and his lord, the king. However, there are several reasons why we shall be able to rule out such an interpretation as unreasonable in the context of the open field system. First, there is the historical fact that the open field system existed side by side with enclosed farms for hundreds of years within the same country. In England, it appears that the open field system was never introduced in Kent or in Cornwall, and there are areas in other counties that also seem to have been enclosed from the beginning of written records.[16] If the open field system were inefficient from an allocation standpoint but efficient as an instrument for income reallocation for a dominating group, then it becomes inexplicable why the system was never introduced in all agricultural regions. Further, the open field system is confined to northern Europe and was never employed in the Mediterranean basin. This points to the fact that the open field system was most probably dependent upon such geographical conditions as soil quality, climate, erosion, that determined the mix of output suitable for open field farming. Thus, we should look for the keys to an understanding of the system in resource endowments,

---

[16] See the discussions of field systems in various regions contained in A. R. H. Baker and R. A. Butlin (eds.), *Studies of Field Systems in the British Isles*, Cambridge, 1973, for a justification of this assertion.

D

technology, and markets, rather than in a presumption that the system was designed to be a vehicle for income redistribution.

Secondly, there is the question of the long-term stability of the system. It remained remarkably intact for hundreds of years, in many diverse cultures and locations. The outstanding feature of the open field system seems to be its resiliency. It weathers many storms, and only drastic exogenous changes will topple the system. This historical stability across time and culture argues very strongly against an interpretation of the system as a design for income redistribution. For it is very simple to show that the statement 'inefficient from an allocation point of view but efficient from an income redistribution point of view' would imply that the system should be expected to be extremely *unstable*. The reason is the standard proof in welfare economics that the allocation of resources and the distribution of the proceeds are separable issues altogether. In the context, this means that if there is an exploiting class that 'sets the rules of the game' in their own favor, this class of exploiters can do better for themselves if they organize an efficient system for the allocation of resources. They will maximize their rents from the land if they, first, attain an efficient institutional organization that yields an efficient allocation of resources and maximizes total income, and, second, maximize their share in the distribution of that maximum income. The thrust of this argument is that the lords and the king, had they ruled the world, would have had strong incentives to set up an efficient economic organization, again given the constraints of the time and place.

In order to demonstrate this, suppose the lords set up an inefficient institutional environment. This is equivalent to stating that the open field system was inefficient, and that an alternative organization existed that was more efficient by yielding a higher total income for society as a whole. If that were so, then the farmers subjected to the inefficient system would do better by inducing the lords and the king to allow a switch to that better organization. There would naturally be an incentive for the king and lords to accept such an improvement in efficiency, for the increase in total income can be divided between the tenants and the lords so that both are better off. This implies that the idea that the lords would maximize their income if they preserved an inefficient structure is a logically inconsistent statement : as long as income distribution and allocation are separable issues, it is possible to show that a Pareto efficient change is possible so that all parties would benefit from establishing an efficient institutional framework.

This is not to deny that the idea of inefficient institutions is a relevant one, nor that cases may exist in which it is indeed possible to show that some class may gain by using the structure in its favor. However, the point of our argument is that such examples ought to be of limited duration and not widely dispersed geographically and/or culturally. We would have to interpret them as disequilibrium situations, and should ask about what constraints in the particular historical context made such a disequilibrium persist for a period of time. Such an approach is completely out of the question in the context of the open field system, for the reasons stated above : it existed for a long time in very different places. In such a situation the argument of disequilibrium is impossible to take seriously. However, we shall show how in the context of enclosure the idea of certain individuals or classes using the structure in their own favor has obvious relevance. In that context, the disequilibrium argument fits in very well, for the process of attaining an enclosure is indeed a state of disequilibrium in that the institutional framework is in the process of moving from one equilibrium to another.

It may have been noted how the previously cited examples of inefficient institutions and property rights assignments pertain to decisions made by the government, either in the form of a congress or a parliament, or in the shape of various regulatory agencies or executive departments. The reason may be sought in the division referred to earlier in this chapter between the two sets of decisions that any economic system must make : the basic one of determining what rights to define, and the ensuing one of voluntarily trading such agreements between various agents who take the first set of decisions as exogenous to their trading agreements. It is clear that the firm is established through contractual arrangements between trading agents who face given property rights constraints. Consequently, it is a decision that is open to various forms of competition. Should the firm indeed constitute an efficient institution, a case can be made for expecting that the trading arrangements made by economic agents will superimplant another institution in its stead. However, owing to the more limited degree of competition involved in the making of the more basic decisions of what property rights to impose on the system, it can reasonably be expected that inefficiencies may result from the influence of various groups that use the political machinery in their own favor. It is consequently not to be wondered at that the examples of inefficient institutions will mostly be found in actions undertaken by the government in the process of attenuating property or decisionmaking rights; for it might reasonably also be expected that, if transaction costs are or can be made sufficiently low, the endogenously created economic institutions set up by individual traders will attain a relatively high degree of efficiency. They would otherwise not survive. This statement should not be taken to imply that the government is incapable of establishing efficient institutions – clearly, the discussion of the efficiency of property rights earlier in this chapter is an example of how the government can contribute to the elimination of externalities and thereby contribute to economic efficiency. However, it does provide us with an additional rationale for rejecting the proposition that the open field system was imposed on the tenants as a method of income redistribution. Since a village always could abolish the open field system through a communal decision, there was never any guarantee that the system would be preserved. Hence, since a viable alternative choice always existed for the members of an open field village, we may reject the idea that the system was dictated by some higher authority.

4

THE ECONOMICS OF COMMONS, OPEN FIELDS, AND SCATTERED STRIPS

On the title page of one of his most renowned works,[1] an eminent historian has placed the following quotation from William Blake : 'To generalize is to be an idiot. To particularize is the alone distinction of merit.' Even among contemporary historians, this general attitude towards historical research does not seem to be uncommon. Consequently, since we are about to present an abstract, theoretical model of the open field system, an apology or justification for the approach might be thought in order, for a great many toes will perhaps feel sorely stepped on. However, none will be given, nor is one really necessary. For what is to be understood by 'generalizing' or 'particularizing' is, of course, a matter of point of reference. What the historian had in mind when selecting the quotation from Blake as a motto was quite probably that, in historical research, general models are easy to come by, but exact explanations for the many variations of history are difficult to achieve. That may be the historian's way of looking at the problem, but it is not necessarily that of the economist. Unless, he would say, there is a theoretical model sufficiently precise as to explain, in a consistent manner, the general phenomenon, various accounts for particular events have not been shown to be relevant, for there is no common theme that has been understood so that the empirical 'particularization' can be justified. From this point of view, it is clear that what will be attempted in this chapter is really nothing but a theoretical particularization : a specific, well defined model will be presented that will account for all the stylized facts of the representative open field village. What is being avoided is the general approach of the old historical school, where every individual phenomenon is more important than the totality of relations formed by the

[1] W. G. Hoskins, *Provincial England*, London, 1963.

interaction of many phenomena – an approach so general as to be atheoretical. Hence, what follows may to an historian look like idiotic generalization, but will, it is hoped, to an economist appear as meritorious particularization – and the conclusions of both may be equally justifiable, for their points of reference, as given by the methodology of their respective discipline, will perhaps lead them to different conclusions about the relative merits of the approach attempted here. This is not to belittle either method: they both have their place, and the intricacies of either should not be underestimated.

For historians and economists alike, there is one formidable 'psychological' obstacle to construing models of the past: it is almost universally believed that past economic institutions were 'inefficient', and that this is the ultimate reason for their disappearance. As a theory of institutions and institutional change, this leaves much to be desired. It is a truism that not only present day technology of production, but also modern institutions, represent developments of ideas and traditions inherited from the past, and it is perhaps also true that institutions, as well as technogoly, have seen a period of advance that stretches back over a considerable period of time. But to make out of this a theory that states that the past was inefficient because it did not organize production or exchange in a way that we have learnt is superior is to render any theoretical explanation for past events pointless and empty. As a consequence, it is no wonder that such viewpoints are coming under heavy attack. Contemporary methodology seems instead to take the preferable tack of asking what constraints actually made the past and its discarded solutions efficient for their particular circumstances.

From a modern standpoint, there are two outstanding features of the open field system of agriculture that seem peculiar and anomalous. The scattered holdings of the typical peasant would seem a costly and inefficient way of arable husbandry. Furthermore, we have been able to show how received economic theory is unanimous in explaining how communal ownership will lead to dissipation of rent and suboptimal resource allocation.

Consequently, in the light of accepted economic doctrine, it would therefore seem that the commons should be divided up into individual private holdings, each controlled by one farmer. For, with collective rights, there would be the incentive for each farmer to overstock, or alternatively the necessity for the community to devise some collective decisionmaking institution, at a certain cost to everyone involved, to decide on the allocation of common rights. In addition, the cost of policing each farmer to ensure the proper utilization of the communal grazing areas would also have to be incurred. Individual ownership rights would save on such costs.

These predictions of economic theory stand, as has already been pointed out, in stark contrast with the factual situation in the open field economy. Collective property rights and communal decisionmaking formed a very important part of the productive process in the open field system. It cannot be argued that land was not scarce in the open field villages. The communal grazing rights were stinted and the courts went to a great deal of trouble to see to it that each individual did not over-use the land for grazing purposes. Nor can it be argued that the collective rights of the open field system constituted the result of a badly performed optimization procedure. The system prevailed for roughly a thousand years over vast areas of northern Europe. If some other institutional structure had in fact represented a better way of organizing production, and especially if it were as simple as private property rights in land, it ought to have been implemented somewhere and then spread across the continent in much the same way that the open field system must have proliferated originally. One cannot argue that there was no feasible way of establishing private property rights, since such rights existed in the open fields themselves.

A similar enigma exists with respect to the scattering of the strips. As we have shown, no satisfactory explanation for this rather strange way of holding land has so far been advanced, although this is not for want of attempts by historians and economists. The inevitable implication is that basic principles of prudent husbandry suggest that the farmers ought to have exchanged strips with each other, so that each attained a consolidated block of land that he could work unhampered by near neighbors. The community could thus have rid itself quite simply of all the unnecessary quarrels so prolific in the court records, as well as of the complicated maneuvering of implements and livestock.

Are we to conclude that, since the factual situation of the open field system does not align with received economic doctrine, and with theoretically established principles of husbandry, the system itself must have been inefficient and inconsistent with the best utilization of scarce resources? This question is not rhetorical, as the suggestion has been made in recent literature.[2] However, if we take that line, it will be impossible to explain why the system ever existed, and persisted, at all. The inevitable conclusion would be that it ought never to have existed, at least not for such a long time over such a vast area, and in such relatively diverse cultures as the Slavic, Germanic, Gallic, Anglo-Saxon, and Celtic. A simple application of existing theory will, it seems, explain nothing and take us nowhere, and cannot account for the shape the structure took.

In addition, there is the complexity of enclosures. In view of the seemingly foolish scattering of the strips and the inefficient joint ownership of the commons, the conclusion that seems firmly established in the literature therefore is that it is not to be wondered at that the open field system disappeared in the enclosure movements from around 1500 to 1800 or so. The only reason for surprise is that it lasted so long and took so long to be replaced by the modern system of farming. Since medieval man can scarcely be thought of as a profit maximizer or even a rational decisionmaker, the explanation goes, we can explain the persistence of the system up to modern times and its disappearance so late by invoking a change in attitudes: rational economic calculations did not make their inroad to agriculture until modern times, and that meant the demise of the old system.

From a methodological standpoint, it is striking how anomalous this explanation for the enclosure of open fields is. It existed for perhaps a millennium, all across northern Europe in a virtually identical manner, and this only because farmers did not act rationally. As soon as they realized their folly, they quickly disbanded the system. Implicit in this kind of argument is a theory of historical evolution that makes changes in tastes and attitudes the exogenous disturbance that yields social changes. However,

---

[2] See, B. O. Baack and R. P. Thomas, 'The enclosure movement and the supply of labour during the industrial revolution', *Journal of European Economic History*, vol. III, 1972, p. 401 et seq.

as pointed out in Chapter 1, this is not a theory, but a tautology, for it yields no falsifiable empirical implications, and is consistent with any observed changes in the past. It is wholly unsatisfactory as a theory of institutions and institutional change, for it makes unidentifiable and unobservable changes in human intelligence the *primus motor* of history.

A preferable methodological starting point might be to argue that, since there seem to be few wanton changes in human nature, there are no changes in unobservable attitudes, but only changes in the given conditions under which human affairs at any point in time are conducted – changes in such conditions, that is, that can be observed and measured. The inference would then be that collective ownership of the commons as a scarce productive resource and scattering of the strips in fact constituted the optimal organization under some given technological, market and transaction cost constraints, and we are thus left with the query as to what exactly those constraints were. This is the approach that will be taken here. Our theory will maintain that, under certain conditions, collective ownership of a scarce resource is a predictable outcome of individual wealth maximizing behavior, and that, under certain conditions, scattering of arable land is quite consistent with intelligent, rational behavior.

The model to be presented is purely static. Therefore, we shall proceed as if it were possible to identify a well defined set of 'institutional arrangements' from which the agents in the economic game will choose the most efficient element. This is not to suggest that the village members in the beginning of time sat down together and collectively made a choice among a large variety of institutional arrangements, the consequences of which they foresaw with perfect clarity, and decided unanimously to institute the best of them all. Yet, for the sake of simplicity and clarity, this is how we will present it here. This static choice construction ought simply to be regarded as a short-cut way of representing a solution that would be arrived at by a trial and error process which probably took centuries to go through. Perhaps no one describes this dynamic process better than Joan Thirsk :

Field systems in their beginnings were almost certainly as varied as villages were numerous. But similar influences tended to iron out diversity and to bring them gradually into closer conformity with one another. Thus, some settlements may have originated as servile communities under the watchful eye of an ever-present lord; others were free associations of men acknowledging none but the loosest ties of dependence. Yet all settlements which developed into villages, and did not dwindle into single farms or disappear from the map altogether, came eventually to consist of all classes of men, holding intermingled lands. Many economic and social trends exerted pressure on them all. All, for example, had to accept the partitioning of land and encroachment on the waste as numbers rose; all could consolidate parcels, or abandon land as demand fell. All were driven to rationalise their use of land as the markets for agricultural produce expanded. . . . The likenesses of many field systems do not prove their origin in a single plan but denote rather the unifying influences which shaped their later development.[3]

Thus, we shall concentrate on this 'unifying influence', and abstract from the process by which a village arrived at a mature system.

The preceding chapters will have made it clear that we shall scorn explanations of the open field system that imply the 'dumb peasant' sheepishly fumbling his way through history, not thinking nor caring what he is doing, until he suddenly hits upon a bright idea like enclosure; yet we must equally steer clear of painting a picture of farmers 'choosing' the wealth maximizing solution from the total set of feasible institutional arrangements. In fact, within their given historical context, agents may be 'culturally blind' to most of the alternatives open to them – their experience may simply not allow them much of a choice. However, it is permissible to ask the more limited question of why certain institutions were preferred over some *known* alternative. This is the problem that will be examined in the following discussion.

Specifically, we shall compare only *two* alternative solutions. The traditional argument has been that the open field system was inefficient relative to enclosed farms. The implication is that historians have come to regard these as two feasible alternatives – a view that would not appear unreasonable as enclosed farms

[3] Joan Thirsk, 'Field systems of the east midlands', in A. R. H. Baker and R. A. Butlin (eds.), *Studies of Field Systems in the British Isles*, Cambridge, 1973, p. 234.

and open field townships existed side by side for as far back as written records. The model presented below should not be taken to imply that farmers in the Middle Ages constantly compared a multitude of alternative solutions. This they did most certainly not do. However, there can be little doubt that they constantly argued over enclosure. If they did not enclose, the implication is that they chose to preserve the open field system, and it is this choice process that the model here is intended to highlight.

Consequently, it should be carefully noted that this is a much more limited notion of efficiency than is commonly associated with that term. The argument below will emphatically not be that the open field system was 'the best of all possible worlds'. Since the set 'all possible worlds' is impossible to define in any relevant manner, or in a manner that would allow medieval peasants to regard all the elements of that set to be available for implementation, we can only make use of a more limited notion of efficiency. The argument presented here will, therefore, only be that the open field system with its specific property rights mixture was efficient relative to the modern system of farming with one man, one owner, one decisionmaker; that is, relative to enclosed farms. We shall hence leave open the question of whether both the open field system and enclosed farms in severalty were inefficient relative to some other form of agricultural organization.

We assume wealth maximizing behavior as the norm for comparing the two relevant alternatives. This must be interpreted to mean *private* wealth maximization. We shall look at the agents in the open field system as pursuing their own personal goals, as if the rules and regulations they set up are the result of joint individual decisionmaking. That is to say, if certain restrictions on individual choice or infringements on personal freedom are observed, we shall interpret these as if they are voluntarily agreed upon by a collective of private decisionmakers, and not imposed upon them against their will by some dictatorial power or by forces of cultural tradition beyond the control of mankind. The thesis here will be that the collective property rights and decisionmaking rules are the result of joint individual optimization and decisionmaking. Later, we shall contrast these propositions with the conclusions reached by Demsetz and Cheung that private ownership in a scarce resource is unambiguously superior to collective, as well as to the more traditional models of the open field system discussed in Chapter 2.

Assumptions and preconditions

In Chapter 3, it was noted that certain conditions must be taken as given when treating property rights and institutions as endogenous choice variables. Notably, these are natural resource endowments, technology of production, and institutional arrangements beyond the control of the village itself. In this section, we shall discuss those conditions, and so set the formal stage for the development of a full theory of the open field village.

The first point to note is that the village constituted an exclusively owned resource, controlled by a closed collective of cultivators. We may express this element by saying that we shall take the village size as given for the analysis at hand. It is convenient for analytical purposes to regard the village as consisting solely of a collective of decisionmaking farmers of a given number, and to take it that the geographical location and distribution of the land in the village are well known. We shall pay no further regard to differences in social status or in land tenure among the farmers in the village. So for the purposes of the ensuing discussion, it is immaterial whether the village under consideration consists of one manor or several, of one village or several, or of free or unfree tenants.

This statement is intended to highlight the very important fact that the productive resources of the village community are under the control of the village members themselves. For example, although the waste and other pasture areas were used communally among the village members, anyone outside the village community who endeavoured to put his beasts out to graze committed trespass on the property of the village.[4] The commons were common only

---

[4] Because of their basic belief in the inefficiency of collectively owned resources, it is economists who make themselves guilty of often assuming that the commons were open to everyone, and that any newcomer to an open field village could partake in grazing and tilling. Historians do not go wrong here – so this assumption is made explicit only for the benefit of readers with training in economic theory. The implication is that the communally owned commons did yield an economic rent, and this is one of the reasons why it is possible to show that they were indeed efficient. On the exclusiveness of the commons, see, e.g., R. H. Tawney, *The Agrarian*

to a certain well defined group of people, and not public property for anyone to use at will.[5] Also, the village members retained control of migration into the village. The ancient system of frank-pledge, whereby some members of the village assumed corporate responsibility for newcomers, guaranteed this state of affairs.[6] Perhaps the most important economic implication of this is that the preconditions existed for the communal waste land still to earn an important economic rent.[7]

Although the statement that the village size is given may appear fairly innocuous, it quite clearly presupposes a well organized institutional structure beyond the control of the village. For example, in case of disagreement between two neighboring villages over the control of woodlands or grazing areas or rivers and ponds, the villages must have had access to the king's courts for the settlement of such disputes. Higher level governmental authority, acting through the king's courts, was ultimately responsible for the delineation of each village. This is consistent with the historical fact that the land in the village was held in freehold by

---

Problem of the Sixteenth Century, London, 1912, p. 238; F. Pollock and F. W. Maitland, *The History of English Law*, Cambridge, 1968, vol. I, pp. 262–3; W. G. Hoskins and L. Dudley Stamp, *The Common Lands of England and Wales*, London, 1963, p. 4.

[5] See e.g., Lord Ernle, *English Farming, Past and Present*, London, 1968, p. 297: 'Both in legal theory and as a historical fact, only the partners in the cultivation of the tillage land were entitled to the pasture rights, which were limited to each individual by the size of his arable holding. Outside this close corporation any persons who turned in stock were trespassers; they encroached, not only on the rights of the owner of the soil, but on the rights of those arable farmers to whom the herbage belonged. Strangers might be able to establish their rights; but the burden of proof lay upon them. Similarly, it was only by long usage that occupiers who rented ancient cottages could exercise pasture rights, unless they also occupied arable land with their houses.'

[6] The frankpledge or the tithing consisted of members who assumed responsibility for each other; thus every man must belong to a frankpledge. Later, the frankpledge became the court leet, where simple infractions against the custom of the manor were punishable and fined. See Pollock and Maitland, *The History of English Law*, vol. I, pp. 568–9, 580; also, J. A. Raftis, *Tenure and Mobility*, Toronto, 1964, *passim*.

[7] This rent could be dissipated if entry into the village was free, i.e., if the size of the village could not be taken as given. For an analysis of this aspect, see S. N. S. Cheung, 'The structure of a contract and the theory of a non-exclusive resource', *Journal of Law and Economics*, vol. XIII, April 1970, pp. 49–70.

someone or other. The lord held his land in fee simple, and whatever land in the village he did not hold in fee simple was held by some other freeholder. With such estates went commons appendant: in other words, the king's courts, which were responsible for the protection of freehold tenure, recognized the rights of the freeholders in neighboring grazing grounds, woods, and waterways, and settled disputes over such matters.[8] These are questions of institutions and property rights beyond the control of any one village, and may be taken to represent the constitutional stage from the viewpoint of the township members.

The second element to be brought out here is the farmers' private ownership rights in the land used for arable purposes, as long as it was used for cropping rather than for the grazing of animals.[9] Again, it is not necessary to make any distinction here between tenants and freeholders. The farmer, whether freeholder or copyholder, decided together with the others on the uses to which the land should be put, such as crop rotations, dates for communal grazing, and so on. In addition, he would decide for himself on detailed matters of cultivation consistent with the overall control of the village community, and had the exclusive right to the crop from his strips. Further, the farmer could rent out the land or rent other lands to add to his holdings. It has already been pointed out that the farmers engaged in quite frequent exchanges of land, subject to the payment of a fee to the court, so it appears quite reasonable to regard even the copyhold tenants as exercising virtually the same control over the land as the freeholders did.[10]

---

[8] Pollock and Maitland, *The History of English Law*, vol. I, p. 607.

[9] Above we have already pointed out the rather interesting feature that the same plot of land could revert from private to collective ownership in a strictly controlled cycle: the strips were owned by individuals, but used communally, without any individual being able to prevent this, whenever the arable was used for grazing.

[10] 'Free peasants had no constraints on their liberty to alienate or sell their land as they please. The earliest charters conveying these lesser estates begin in the twelfth century and from the outset show a flourishing market in land and a willingness to carve up holdings in all manner of ways in order to convey portions as gifts, exchanges, or sales. And the charters only show the tip of the iceberg, since a host of transactions continued to take place by livery of seisin without written record. All these had the effect of dividing some holdings and consolidating others. Both purposes were promoted by the market in small parcels of land.' Joan Thirsk, *Field Systems*, pp. 267–8.

We shall use the term private property rights to represent the status of both the copyhold or villein tenant and the freehold tenant. This does not imply that either of them should be regarded as a private owner in the modern sense of that word – clearly, such an interpretation would be historically unjustifiable. Nor is the use of the phrase 'private property rights' intended to imply that the legal interpretation of the Middle Ages was to grant ownership rights to tenants; again, that would not be a justifiable interpretation. Rather, the phrase should here be taken as a well defined technical term in economics (as apart from legal history) that implies only that certain well defined rights of decisionmaking were vested in the individual tenant. The distinction between freehold tenant, copyhold tenant, and a lord of the manor is then not so much in the *kind* of decisions they each made, but more perhaps in the amount of economic resources they controlled. No one will ever deny that this is an important distinction, and, indeed, we shall discuss some of the ramifications of income distribution problems below. However, from the point of view of understanding why the open field system was preserved when enclosed farms were a viable alternative, it is not relevant or necessary to make a distinction between various kinds of tenure. An additional justification for this simplification is provided by the historical fact that by no means all open field villages were manors. In many areas, open field villages existed that were collections of freeholders only, and it would therefore be a mistake to argue that the relationship between a lord and his tenants is the focal point of the open field system.

This simplification allows us to slur over an important but extremely difficult historical evolutionary process: the disappearance of the villein, and the emergence of the copyholder. It seems fairly clear, as clear as anything in feudal legal theory, that the villein tenant never was the owner of the land he tilled. The land 'belonged' to the lord inasmuch as it was held in fee simple by him, and the lord would work his demesne with boonworks. The villein owed corvée to his lord, and in return he obtained the right to till a certain amount of land. The land rights of the villein must be regarded as a payment for the labor he performed for the lord. In other words, in the initial stages of the feudal period, the villein was not the owner of the land. Later, corvée and other payments were commuted into a money rent: when the villein emerges as the copyholder of a later time he is clearly paying for the use of the land – not receiving the land as payment for labor performed. The nature of the contract has changed although the exchange itself appears identical. At the same time the copyholder has established some rights in the land, an inevitable tendency in a country such as Britain where the common law is such a strong concept. In the end, it would appear that the only right that the lord retains is the right to collect certain rents and fees from the land.

Why did this change from an initial situation where the lord managed his farm with hired labor to a later one where the farmer controls his own land ever occur? For an economist, the simple answer would appear to lie in the incentive structure afforded by the two alternatives: an owner has something at stake in the use of the land, and will therefore tend, on average, to be more concerned about its efficient use than a hired hand. This can account for the decline of the demesne, its renting out to tenants and for the pervasiveness of copyhold tenure. For our present purposes this simple answer will have to be accepted.

The third feature of the open field system that we shall stress is the production of livestock and grain in all villages. For simplicity, we shall proceed to characterize the open field economy as a system of production of only these two interrelated outputs. This is not to deny the existence of a rich diversity of crops, nor the fact that animals of many kinds were reared. For the purposes of the analysis here it is sufficient to regard wheat, oats, peas, barley, hemp, flax or turnips as one homogenous crop, which we shall choose to call grain. Similarly, it is not necessary for the ensuing argument to make any distinction between sheep, cattle, horses, pigs, ducks or chickens – it is sufficient to refer to them simply as livestock. Even with this simplification it is a rather complicated system of production, for each of the two outputs will require treatment as an input into the production of the other. Apart from land and labor, grain production employs livestock for ploughing and carting, and their manure for fertilization; and livestock production employs grain for feed. This is not necessarily to say that any of the white crops was used for feeding cattle or sheep during winter. With our simplified treatment of the outputs, hay and turnips are regarded as 'grain' as well, and, at least in the later period, turnips were useful for feed. This interdependency between animals and crops makes for a very complex choice of how much of the given land area to put under plough, and how much to leave for pure grazing. The ratio between the two was never fixed, but varied with local conditions, with changes in population, with changes in production techniques, and with relative output prices. Postan expresses clearly the complexity involved in the determination of the amounts to be produced of the two outputs :

The areas under corn could grow only at the expense of the areas under grass, and most of the reclamations necessitated the taking in of communal and manorial wastes, previously used for pasture, however rough. The resulting reductions in pasture were bound to affect the entire economy of villages. Presumably, they were felt least of all in the purely pastoral areas of England, among her uplands, marshes and forests, given over exclusively or mainly to cattle and sheep. But in areas where mixed farming prevailed and where men depended for their living on the grain they grew – and such areas yielded most of the medieval England's produce and carried most of her population – the continuous reduction of pasture could threaten the viability of arable cultivation itself. The threat was inherent in the very nature of mixed husbandry. In the latter the pastoral element not only provided a direct source of goods and income but also served certain essential needs of grain growing. Some grass was needed to maintain the animals employed in ploughing and cartage; but in addition arable farms needed animals as their only source of manure.[11]

Because of this nature of the relationship between the outputs, complete specialization in the production of one of them was virtually impossible, and for the purposes of the following discussion we shall rule out such pure specialization in the context of the open field system. There were areas of England where agricultural production tended to specialize in one of the two outputs, but the specialization was never complete. Furthermore, those areas which tended to concentrate on the production of one of the outputs did not have the open field system.[12] The open field

[11] M. M. Postan, *The Medieval Economy and Society*, Berkeley and Los Angeles, 1972, p. 57.

[12] For some areas in England that specialized in the production of one of the two outputs considered here at an early time it is extremely doubtful

system was characteristic of the areas of mixed farming where both outputs formed integral parts of the income the farmers realized.[13]

Yet the assertion that both animals and grain were produced in all open field villages is not simply intended to capture the technological interrelationship between them, but to stress the importance of the limited size of output markets. England is a country of very varied soil patterns and qualities.[14] Such differences would naturally give rise to regional comparative advantages in the production of some specific output – grain or cattle in our simple terminology, or, in a framework richer in crops and animals, in white crops, turnips, legumes, etc., or in wool, mutton, dairying, fattening, etc., as the case might be. However, in the open field system, probably due to a costly and inefficient transportation system and to the lack of large non-agricultural population centra, trading was cheaper in inputs, most notably labor, than in outputs. As a consequence, concentration (as distinct from complete specialization) in production was barred, and both outputs were produced in all open field villages. The Orwins also express this view :

It is no more than a fair generalization, too, to say that farming in

---

whether the open field system was ever in use at all. In Kent, for example, the arable fields were laid out on a pattern of rectangular fields, with no scattered strips and no communal grazing, as opposed to the irregularly shaped open fields in other areas. A very good analysis of the Kentish system is given by A. R. H. Baker, 'Field systems of southeast England', in Baker and Butlin, *Studies of Field Systems*, pp. 377–430. Also, see Joan Thirsk, *The Agrarian History of England and Wales, 1500–1640*, vol. IV Cambridge, 1967, p. 9, and M. M. Postan, *The Medieval Economy*, p. 52.

[13] See e.g., Joan Thirsk, *Agrarian History*, vol. IV, p. 197. '. . . pastoral areas were engaged either in cattle or sheep breeding, fattening, dairying, pig-keeping, or horse-breeding, or, more usually, doing something of each. . . . Always the arable land in these regions was no more than adequate to satisfy domestic and farm needs for grain, and the large-scale market production of arable crops was impossible. The mixed farming areas usually contained far more ploughland than pasture. . . . Mixed farming regions specialized in growing grain and fodder crops. Cereals were marketed, if water transport by river or sea was available, and if not, were used to feed and fatten the stock for the butcher.' Most of the open field areas were in the category of mixing arable with fattening, i.e., marketing both outputs.

[14] E. Kerridge, *The Agricultural Revolution*, London, 1967, recognizes some forty major farming regions of England alone, and this only at the expense of some generalization.

the open fields was practised for self-supply. The isolation of village communities for a large part of the year threw them entirely upon their own resources. Even so, there was a certain amount of trading within them, between the village craftsmen and the husband-men, who together made up balanced societies. Beyond this, farming for the market was restricted to those places within reach of the towns. For most of the inland countries, this meant a distance of comparatively few miles. Right down to the eighteenth century, heavy traffic was impossible on the green roads which formed the bulk of the highways, and pack-horse transport had obvious limitations for bulky goods.[15]

This is not an assertion that markets did not exist, nor that they were unimportant. Clearly, such an assumption would be unwarranted in the face of all the trading that actually occurred in the open field villages, both between its members and with outsiders. The assertion is one of size of markets, i.e., we should expect that trading was typically localized to the immediate vicinity of the open field village.[16] This point is often stressed in the literature, and the following quotation from John may serve as an example among many:

A more fundamental reason lay in the inadequacies of contemporary means of transport which accentuated the effects both of regional markets and of regional climatic differences. Metropolitan and export demand dominated the corn trade of the Thames valley and of the Eastern districts; and the south-western area might well have been divided into two groups of markets, one based on Bristol and the other on the inland clothing towns. It is probable that, because of

---

[15] C. S. and C. S. Orwin, *The Open Fields*, Oxford, 1967, p. 24.

[16] In his paper, 'Champion and woodland Norfolk: the development of regional differences', *Journal of European Economic History*, vol. 6, no. 1, 1977, S. Yonekawa shows a map of Norfolk at the end of the fourteenth century. The county contained 89 towns that held weekly markets (p. 166). Thus, there is practically no point, except in the southwest corner and around Norwich, a larger population center, where a farmer would be more than a couple of miles away from a market. Another writer notes: 'Scattered up and down the countryside, at intervals of every few miles, there were about 760 market towns in Tudor and Stuart England and 50 in Wales, each with its official weekly market day or days and its fairs held once or twice a year. . . . Despite this proliferation of markets, there were far fewer market towns and villages in the sixteenth century than three centuries earlier – probably less than one third as many.' A. Everitt, 'The marketing of agricultural produce', ch. VIII in Joan Thirsk (ed.), *Agrarian History*, vol. IV, p. 467.

growing industrial concentrations, the area denominated 'north
of the Trent' might also have contained not one, but several
markets.[17]

The farmers in open field villages obtained a significant proportion of their income from the sale of both livestock and grain.[18]
Our observations that both outputs were produced in all villages
and that output markets were small are intended to focus attention on this feature, for it is a remarkable one when related to the
varied soil conditions of England and the concomitant potential
comparative advantages in different areas. As long as small output
markets effectively prevented regional specialization and inter-
regional trade – not necessarily as between specialized villages
importing one of the two commodities for its own consumption,
but between different regions producing for sale to non-agricultural population centres – these comparative advantages remained
unexploited.

The fourth feature of the open field system that we shall
emphasize is the universal practice of rotating husbandry in the
arable. Although the introduction of convertible or rotating husbandry is often thought of as belonging to a later period, it is a
simple historical fact that it was practiced in the open field system
for almost as long as records exist. Depending on whether a two-
field or three-field system was used, half or a third of the arable
would lie in fallow each year. The reason for this custom is commonly asserted to be simple agricultural prudence: with more
intensive cropping the soil would become exhausted, and a period
of rest would be necessary to avoid declining yields. Whatever
the reasons for the practice of rotating husbandry in the open
field system, it seems clear that fallowing and grazing on the
fallow was a universal feature of the areas of champion fields.

Soil exhaustion or not, the question of whether to employ
rotating husbandry hinges on all those factors which determine
the intensity of the use of the soil that the farmers settle on as

---

[17] A. H. John, 'The course of agricultural change', in L. S. Pressnell (ed.),
Studies in the Industrial Revolution, Oxford, 1960, pp. 126–7. Postan,
The Medieval Economy, pp. 197, 207, also makes the same point.

[18] 'Few farmers were so self-sufficient on their farms that they could ignore the
market entirely.' Joan Thirsk, *Agrarian History*, vol. IV, p. 3.

the best in their particular circumstances, as these are revealed to them through a long and extended trial and error process. Among the conditions that determine this choice will be the available agricultural techniques, soil condition, relative output prices, and demand for the final products. The problem is not only to determine crop rotations, but also to decide on the proper combination of livestock and grain.[19] Although these two outputs, as has been pointed out above, each require the other as an input in proportions that over the long run are variable only within narrow limits, it is possible, in the short run, to increase production of one at the sacrifice of the other.

During the Middle Ages, fallowing and grazing on the fallow probably was the only technique for rotating husbandry known. Later, in the sixteenth and seventeenth centuries, the practices known as up-and-down husbandry and floating of watermeadows were developed. These methods are designed to increase the intensity of the use of the land; yet even with these developments the grazing of fallow land was never abandoned in the open field villages. However, in spite of the fact that grazing fallow land may seem an intelligent and economical use of land, it is not as self-evident a practice as it may appear at first.

The simple point is that rotating husbandry is not a costless practice even in the fallow, although it may seem that the land temporarily has no alternative use. For, in order to employ the fallow for grazing, it was necessary to impose restrictions on the farmers of the village to ensure that an extended area for grazing could be obtained. This must have meant that the freedom of the farmers to follow any crop rotation they pleased was restricted, as indeed many bylaws give example of, although special permission to deviate from the usual rotations would sometimes be granted. This is costly not only in terms of bargaining, negotiating, and policing agreements, but also in terms of output, if actual crop rotations were different from those desired on the part of at least some farmers. Rotating husbandry in the open field system must have severely restricted the freedom of the farmers, with

---

[19] As population grew, the pasture was gradually invaded by the arable. Population pressure could thus upset the precarious balance between the two. M. M. Postan, chapter on England, in Postan (ed.), *The Cambridge Economic History of Europe*, vol. I, Cambridge, 2nd ed., 1966, pp. 553–4.

resulting losses of output of some types of grain. But it is not clear that simply grazing the fallow adds any nutrients to the soil. If the animals stay confined to the same area, they will not return more chemical elements through their spillings than they remove by their feeding, so the net addition will be zero unless the animals are put to grazing on some other plot and then moved to the fallow in order to capture the spillings that result from feeding them elsewhere. This was done in the open field villages. Special foldcourses were constructed where the animals were herded at night. Again, this is not a costless procedure as the animals have to be moved back and forth, and fences erected. Even in early medieval times alternative or complementary means of fertilizing the soil were known, such as liming and marling, and dung was collected and carted out on the fields.

The three-field system is considered a later development of the two-field system, and is hailed as a great innovation by some writers : rather than having half the arable in fallow each year, only one third is uncultivated at any point in time. There can be no doubt that the output from the arable increased greatly by this change in techniques, yet the inference cannot automatically be drawn that total income increased by the same amount as the increase in crops. For the increase in the use of the soil for arable necessarily meant a decrease in the amount of pasture available : the three-field system must have resulted in less output of animals. And if the number of animals was reduced, the income of the village as a whole may not have increased as much as would be inferred from simply looking at the increase in the acreage under the plough. This is to stress the obvious : rotating husbandry represents an intensive use of the soil, and the three-field system is much more intensive than the two-field system. Such an intensive use of the soil would only be profitable where the amount of people to be fed off the land is great, and where market conditions and relative prices are such that it is desirable to produce both outputs as cash crops.

Common grazing on the fallow and common of shack were not universal features of English agriculture. In the hilly uplands, where the farmers specialized in livestock production, the topological conditions were such that the rather small areas of arable were not collected into great champion fields with collective grazing.[20] There was abundant pasture available on the commons. Often these pastures were not even stinted, and newcomers could be allowed to put out their beasts without much ado. This was never the case in the areas of the open field system. There, grazing was a valuable right, jealously guarded.[21] The example of the pastoral areas of England shows that common grazing on the arable is not costless to achieve, for these areas never organized their communities around rotating husbandry. The same is true of the areas of Kent where the open field system never existed. Here the farmers specialized in the production of grain, and communal grazing was never universal.

The great significance of rotating husbandry is that it points to two important features of the open field system. First, the system was geared towards very intensive use of the soil. Presumably this would coincide with the areas that were densely populated as well – we shall leave it open which way the causation runs, whether from good land to many people, or from growing numbers demanding more of the soil.[22] Secondly, the open field system was geared towards achieving the maximum long run output of *both* products, grain and livestock, for otherwise – as was the case in the areas which specialized early in one of the outputs – the practice of rotating husbandry on the basis of a communal organization would not be necessary.

One more assumption must be introduced before the analysis of the open field system can get properly under way. This relates to the grazing practices in the open fields and on the commons. Grazing was communal, on a large scale, and we must therefore assume that the optimal pasture area was large. This is a key assumption in the theory to be developed. It is equivalent to

20 'In the pastoral districts of England, the more typical unit of settlement was either the hamlet or the single farmstead having little working association with its neighbors except sometimes in the use of common grazing grounds. Manorial control was more difficult to exercise since the centres of settlement were many, and farming matters demanding communal regulation were so few as to afford little occasion for bringing the community together.' Joan Thirsk, *Agrarian History*, vol. IV, p. 9.

21 *Ibid.*, p. 12 asserts that unstinted pasture was not to be found in lowland England by the beginning of the sixteenth century.

22 Joan Thirsk will be found to argue that the causation runs both ways. From land to labor: *Agrarian History*, vol. IV, pp. 13–14; from labor to land: 'The common fields', *Past and Present*, no. 29, December, 1964, pp. 8–9.

asserting that, for a given number of livestock and a given amount of land divided into given areas of pasture and arable, the yield from the animals would be greater if the herd is permitted to graze together on one large plot of land rather than being split up into smaller units, each with its own small plot to feed on.[23] This would be the case either if the total product from the herd, for given costs of production, is greater if it is kept together in one unit, or if the costs of achieving a certain output were reduced if the herd is kept together.

It is perhaps difficult to argue that the yield of a given number of animals would increase if they are kept together rather than divided into small units – a rather esoteric argument of the kind 'cattle are a gregarious input, and happy cows produce more' would be required. Although there may be some merit to this, we shall not rely on such an argument here. However, it is well known that the grazing habits of cattle are such that they naturally stray while eating, so a large area would be more conducive to accommodate such habits than several smaller ones. Indeed, this habit of cattle is so important that it was the very foundation of the cattle trails in the old American West. On the trail the cattle were not driven at all, since that would have made them arrive at the market place much too thin, but only pointed in the desired direction and permitted to graze as normal. They will thus travel about eight to ten miles a day. Therefore, if the herd is kept together, the grazing becomes more extensive as the animals wander about and allows the grass in any one area to recuperate before it is trampled down again by the grazing animals. Thus more grazing may be had out of a given area if the herd is kept together rather than split up into smaller units.

It would not be prudent to make too much out of this argument. Perhaps the more important reason for accepting the proposition that the net yield is higher for a large flock than for

---

[23] The exact nature of the assumption may be stated as follows :

$$\left| \frac{\delta Q_a / Q_a}{\delta N_a / N_a} \right| < \left| \frac{\delta Q_p / Q_p}{\delta N_p / N_p} \right| \quad A = A^\circ, \ X = X^\circ,$$

where Q is output, N number of plots, the subscripts a and p refer to arable and pasture, respectively. The elasticity must be taken for given total area (A°) and given number of animals (X°). Both partials are negative, hence the absolute values.

several small ones lies on the cost side. It is quite clear that the costs of supervising a herd of cattle or a flock of sheep are subject to substantial economies of scale : a shepherd and a dog can guard a substantial flock consisting of the animals from several owners. The village would collectively employ a shepherd and a herds-man, paid for jointly by the owners of the animals :

Each morning during the summer the village shepherd, assisted by his dogs, drove his sheep to the pastures, keeping a watch upon them all day to see they came to no harm. Twice a day (three times a day in the month of May) the ewes were led to the sheds to be milked, there being some demand for ewes' milk, butter and cheese. At night the sheep returned home, and were folded on the fallow in a 'common fold', which was moved at intervals in such a way that the field should be manured more or less evenly. The shepherd slept in a moveable hut near his charges. . . . In summer the [cattle] lived an outdoor life. Each morning the herdsman passed through the village blowing his horn, at which signal the cattle of the different owners not required for field service went out to join the swelling herd, which made its way to the pastures by cattle routes or drift ways fenced with hurdles or fences to prevent straying. The herdsman spent the whole day with the cattle and at night brought them back to the village, where each animal was returned to its own shelter as the diminishing herd went by.[24]

Apart from the costs of supervision, there would be the cost of fencing in the small plots for grazing if the herd was divided into smaller units, and it is very easy to show that the cost of fencing is higher per acre the smaller the number of acres : by keeping the animals in a large flock, the cost of production would be lower per animal if these fencing costs could be avoided. It is quite probable that both these elements were significant in making large herds or flocks economically superior.[25]

It can also be shown that those who specialized in the produc-tion of livestock in certain areas held quite sizeable herds.[26] Tawney supports the contention of increasing returns in live-

[24] E. H. Carrier, *The Pastoral Heritage of Britain*, London, 1936, pp. 71, 73.

[25] Even modern writers, such as Yelling, support this contention of increasing returns to the production of livestock. See J. A. Yelling, *Common Field and Enclosure in England, 1450–1850*, London, 1977, pp. 96, 101.

[26] K. J. Allison, 'Flock management in the sixteenth and seventeenth cen-turies', *The Economic History Review*, 2nd series, vol. XI, 1958–9, p. 99.

stock : 'While in the case of sheep and cattle grazing on the large scale practised by the graziers of the period, there was obviously no question but that an extensive ranch which could be stocked with several thousand beasts, was the type of holding which would pay best.'[27] Carrier also accepts this proposition : 'The commercial sheep farmer owned flocks varying from 5,000 to 24,000 head.'

There is yet another important consideration for why it is reasonable to suggest that there were substantial returns to scale in the management of animals. This has to do with the importance of manure in keeping the fertility of the open fields at an acceptable level. Joan Thirsk states this lucidly :

In all areas of mixed farming, the folding of sheep on the arable was a pillar of the farming system, and the breeds of sheep used for this purpose were specially conditioned to it. In a common-field area the sheep were put into a common village flock, since a small flock was useless, hurdled at night, and the pens moved from one part of the field to another daily. On enclosed farms the farmer needed a large flock or the system did not work. A thousand sheep would fold an acre of common-field land in a night, and folding was arranged so that each field should be dunged in time for sowing.[28]

With this specification of the environment, it is now possible to proceed to what will constitute the object of analysis : the choice of institutional arrangements in the open field system. The problem we shall examine is the one of why (a) the commons (the 'waste') were owned communally, (b) the arable fields consisted of scattered strips collected into larger units called fields, and (c) the village members organized a formal collective decisionmaking procedure in the courts. Throughout, we shall compare the arrangements as they are known to have been construed in the open field villages, with the alternatives presented by private, individual ownership and decisionmaking control, as they were known to have been implemented in the villages after enclosure, and in those farmsteads which were never part of an open field village.

27 Tawney, *The Agrarian Problem*, London, 1912, p. 214.
28 Joan Thirsk, *The Agrarian History*, vol. IV, p. 188.

The property rights in the waste

The stage is now set: the assumptions are outlined, and the theoretical framework sketched. On the assumption that the arable is of given size and divided into private individual lots, we begin the analysis of the open field system by considering the economic reasons for the choice of property rights in the non-arable lands of the village, i.e., the waste. The 'waste' may not be thought of as 'wasted' in any relevant sense other than that it was not being worked for arable purposes. The waste had important functions in the open field economy. It was the source not only for grazing of the livestock, but for fuel, minerals, timber, water for irrigation and for the mill, and had other uses as well. However, in the simplified framework of our model the waste is considered only as land available for grazing of the livestock. The problem we posit is whether to choose private, individual rights in the waste, whereby the land would be divided up into lots held by individual farmers in severalty, or whether to choose collective rights in the grazing grounds, whereby all the farmers in the village jointly own and control the use of the waste.

Given the size of the arable fields, and given that there are private property rights in the arable, the economic problem becomes one of organizing the ownership of the waste so as to achieve the maximum grazing out of those lands, or, alternatively, to exploit the grazing to be had from the waste as cheaply as possible. Here the assumption of increasing returns to scale in grazing is crucial. We assert that, for a given number of livestock and a given size of the grazing grounds, a greater net yield in terms of animal output is achieved if the animals are permitted to graze together in relatively large herds or flocks on relatively large areas. The reasons for this may be found either in an increasing yield to greater herds, or a decrease in the average cost per animal due to economies of scale in fencing or supervision. The problem therefore becomes whether it is cheaper to exploit these scale economies under private or under collective property rights in the waste. The answer to this question will hinge on the transaction cost structure present under each of the two alternatives, and it is therefore necessary to establish the nature of the relevant transaction costs.

. For the purposes of the present discussion we shall understand with transaction costs those real resource losses that are involved in moving through a process of production and exchange from an initial allocation of resources to the final outcome for consumption. The operational definition will contain four separate elements : the costs of establishing and protecting property rights, the costs of decisionmaking with respect to the use of a scarce resource, the costs of establishing certain organizations to facilitate production and exchange and, lastly, the costs of policing the implementation of the decisions about the desired use of productive resources. We shall show that there are systematic differences between private and collective property rights in relation to these costs as they pertain to the open field system, and that this provides a partial answer to why collective property rights were preferred to private in the grazing grounds of the open field villages.

In our discussion of Demsetz' analysis of the superiority of private property rights in a scarce resource, we made no reference to the costs of establishing property rights. Demsetz, and other writers, proceed as if it were costless to establish – and protect – private property rights. Yet this is not the case, for to establish property rights entails making a decision about who shall have the right to exclude others from the use of the resource. Furthermore, private property rights cannot be said to be established unless they are also protected from intrusion by others who may lay claim to the resource. With respect to the waste it therefore follows that, if it were desirable to establish private property in the grazing areas, it would not be enough simply to stake out boundaries between tenements and assign ownership rights to them, which in itself would be a costly and difficult procedure; it would further be necessary to accept the costs of protecting each owner from the intrusion of others. Alternatively, if the decision were to keep the waste 'common', that is, as collective property of the members of the village community, there would be *no* costs of establishing property rights for individuals in the waste – as long as the membership in the village community was easily controlled. We have seen that the system of frankpledge in the open field villages ensured that the village land was a 'collectively exclusive' resource. As long as the waste was kept common, therefore, the only problem would be to keep members from outside communities out, and this is something the king's courts were designed to handle. Therefore communal property rights in the waste for the village members would at least save on the costs of internal litigation over whose property the waste is, for it would belong to the community as a whole. Communal property in the waste would also save on the costs of staking out and defining individual rights. So, with respect to the first element, the definition and establishment of property rights, it seems reasonable to conclude that collective rights would be considerably cheaper as long as the community is well defined and outside users can be kept at bay to prevent the rent from being dissipated. We already know that this was the case.

Yet this does not clinch the argument, for there are other transaction costs to account for as well. Let us assume, for the moment, that we are contemplating a situation in which the waste has already been divided into private, individual shares. Given that it is desirable to establish a large area of grazing, in order to keep the herd together and realize the returns to scale in grazing, it would then be necessary for these individual owners to reach a joint decision of throwing their lands open to their neighbors so that the large-scale grazing can be achieved. This would require a chain of transactions between the involved parties whereby each individual owner would have to transact with each and every one of the other owners to receive the proper compensation for the use of his land for grazing by others, and for his payment to the others for his use of their lands for his own grazing. The same would be true every time some damage were done to the property of one : he would have to demand compensation from every other user of the land if the damaging party was difficult to identify. If n is the number of owners, it is easily ascertained that the number of transactions involved would be $n(n-1)/2$. Naturally, the costs of such transactions would be substantial. It is therefore safe to conclude that, even if there were private ownership of the waste, some sort of institutional arrangement would be organized whereby such prohibitive transaction costs could be avoided.[29] For example, a decision may be reached to

---

[29] See J. M. Buchanan, *The Demand and Supply of Public Goods*, Chicago, 1968, p. 86, where he says: '. . . individuals will suggest n-person "rules"

abide by the rules of a majority vote : the rents and fees the farmers agree to pay each other for the grazing rights may be determined by some voting procedure rather than by individual bargaining. If compensation were demanded for damages, the same method could easily solve the problem. An institutional arrangement could therefore, it seems, avoid the large number of transactions involved in private ownership of the waste.

Such an organizational device for coping with the inter-relationships between the owners of the grazing animals would of course also be necessary under collective ownership. If the village community, as one body, were the owner of the waste, they would have to determine jointly, by some voting procedure, the best use of the communal waste, just as under private owner-ship. In this case, with respect to the second element in the trans-action costs, it is difficult to see any difference between the two cases of ownership. There can be no presumption that the costs of decisionmaking would be different once we allow for the estab-lishment of an organization to deal with the problem of joint decisionmaking.

Yet there is such a crucial difference, for organizations do not emerge out of thin air. They must be established by transacting agents in an economic environment. Therefore, the problem becomes one of whether it is possible to say something about the costs of establishing and maintaining the institutional arrange-ments, the decisionmaking rule as it were, which would be neces-sary under both collective and private ownership of the waste in order to decide on its use and on the division of the costs. It may at first seem unnecessary to approach this problem at all, for if it is assumed that the costs would at least not be higher under collective property rights than under private to reach agreement on the decisionmaking rules, it would follow that collective rights would be superior to private as long as there are positive costs of establishing private property rights. But we would not want to rest our case on such a tenuous argument. The conclusion can be strongly reinforced by a comparison of the costs of arranging

---

or "arrangements" aimed explicitly at reducing or eliminating the inefficien-cies generated by independent behavior. In a very broad sense, agreements on such rules can also be classified as "trades".' See also pp. 153–8 for a discussion of decisionmaking rules to overcome transaction costs.

the joint decisionmaking procedure. It is possible, that is, to show that such a rule would be costlier to establish under private property rights in the waste than under collective.

Conceptually, under private property in the waste, an agreement on a joint decisionmaking rule, or an institutional arrangement whereby all concerned farmers could influence the use of the waste without having to resort to individual bargaining every time, could only be reached by a voluntary agreement on the part of the individual owners to part with some decisionmaking power over their lands. In order for the joint decisionmaking rule to be effective it would be necessary for the farmers to agree to abide by the decisions as reached by the joint decisionmaking formula. This would entail a partial transfer of property rights from the individual to the collective of farmers, i.e., all farmers would acquire decisionmaking rights over each other's property. It would be desirable for each farmer to agree to this, since the total output, and thereby each individual's share, would be increased by the exploitation of the increasing returns to grazing. However, it is seldom that men's opinions on the most profitable use of a productive resource coincide. There is always the danger in an organization which is voluntarily established that some of the members may decide to withdraw. As long as the institutional arrangement is strictly voluntary, there is no guarantee that the members will continue to play according to the rules and not prefer to go their own way. Yet, if there are increasing returns to scale in grazing, such behavior would impose costs on the other owners of animals. Since it is obviously profitable to stay in the organization, no one individual may actually withdraw – but as long as he can make a credible threat that he might, that individual has acquired a bargaining position against the other members of the organization that he can use for his benefit. By threatening to withdraw and thereby impose costs on the other members, he may be able to appropriate some of their gain from the institutional arrangement. In order to keep some stubborn misfit in line, the others may then have to engage in continual bribing to keep the organization viable.

In this context, it is clear that collective property rights are markedly superior to private with respect to the ownership of the waste. If the waste is owned communally, no individual can withdraw his property from the use by the others – for no individual owns any geographically defined piece of soil. Rather, they all own a servitude which is the right to grazing on a certain piece of property. The property itself is owned by all of them collectively, and is controlled by the collective. If anyone wants to withdraw, this is possible by selling the conditional right to grazing to somebody else. What is not possible is to withdraw the land itself. That is to say, no one individual acquires a strong bargaining position *vis-à-vis* the others if the property is retained in communal ownership. With respect to this third transaction cost, therefore, collective property turns out to be considerably cheaper than private. Setting up and preserving the joint decisionmaking organization becomes easy.

There is yet a fourth cost involved in the choice of property rights in the waste, however, and this has to be taken into account before the final answer to why collective property rights in the waste were preferred to private in the open field system. This is the cost involved in implementing the decisions on the proper use of the waste, the familiar problem of policing that no one imposes costs on the others by his use of the resource; specifically, in the present context, that the grazing grounds do not become overstocked. With collective property rights in the waste, the free rider problem is potentially present in all its force. Each farmer has the incentive to put out more animals on the commons than he would if he were the only owner of the grazing grounds, for the costs of the overgrazing will be borne by all the other members of the community, but he will capture all the increased grazing for himself. Thus, communal ownership will necessarily entail the additional cost of policing that no farmer overstocks the commons. This problem was prevalent in the open field village. There were numerous bylaws to prevent overstocking, and the court records are full of cases of farmers being fined for transgressing these rules.

It might be thought that the establishment of private property rights in the waste would eliminate this problem. Yet, on the approach taken here, this would be an error. Let us suppose, hypothetically, that it were possible to establish private property rights in the waste at zero costs, and that there were no problem with any individual threatening to leave the organization and by extortion appropriating some of the others' income. In this situation of private property, would there be no incentive to overstock the waste? Quite clearly, for each individual farmer, the incentives would be as strong as under collective property. If he could put out more animals than his quota, he would again impose an externality on the others, and appropriate the gain for himself. In this case there is no difference between private property and collective in the waste. What is crucial is not the ownership, but the usership and control. The land might be owned privately, but if it is used and controlled collectively, each farmer has the same incentive to become a free rider as he would under collective property.

The conclusion is therefore unambiguous: if there are private property rights in the arable, if outside non-owners can be kept out, and if each farmer practices mixed husbandry, collective rights in the grazing areas can unambiguously be shown to save on transaction costs as compared with private ownership if there are increasing returns to scale in grazing. This conclusion can only be strengthened if the waste has additional uses which every farmer would want a share in, such as digging peat, gathering firewood, cutting timber, excavating minerals, irrigating meadows, as long as there are some returns to scale in the production of these services.

This explains why the commons were commons and were not parcelled out to owners in severalty. It all hinges on the transaction costs of achieving a large grazing area, given the private rights in the arable, which we have asserted are due to the costs of overseeing hired hands in the fields. Yet the arable fields constitute a puzzle in themselves: it is now time to turn to an analysis of the scattered strips.

A theory of the scattered strips

From an economic standpoint, the collective ownership and control over the commons is only one of the puzzling features of the open field system. Having attempted to explain this particular one by invoking increasing returns to scale in grazing in an environment where mixed husbandry is desirable, and by showing that the costs of attaining those returns would be higher under a

E

system of private ownership in the waste, we shall try to extend this theory to account also for the scattering of the strips within the open fields. But it is also necessary to account for the existence of *open* fields in the first place.

In our simplified exposition of the complex and diverse phenomenon that the open field system constituted, we have been content to represent it as only two or three fields. The fields were usually divided into furlongs, and often the furlong constituted the unit of cultivation in the sense that different furlongs within the same field could be sown with different crops.[30] However, the practice of assembling the strips and the furlongs into fields was a universal aspect of the open field system. To account for this practice, it is again sufficient to point to the desirability of obtaining relatively large areas for the animals to graze on – if it can be shown that rotating husbandry is a desirable method of production, i.e., if it is profitable to alternate the same plot between arable and pasture. There can be no doubt that this was a universal practice in the open field system. The farmers had common rights of grazing on each other's plots in the fallow year, and in most places also common of shack, the right to grazing before ploughing and after harvest. If it is desirable to alternate the use of the land in this fashion, the argument about increasing returns to scale in grazing still holds true, and more net output would be achieved if the grazing could be organized on a relatively large area. This is what the open fields achieve. Each year the farmer will have a certain amount of his lands in fallow, the exact proportion determined by soil conditions and fertilizing methods. This fallow land will be available for grazing, perhaps not throughout the entire year if it is ploughed up once or twice, but there will be a substantial amount of grazing to be had from it. If there are increasing returns to scale in grazing, the farmers will all benefit by collecting their lands into fields so that all the lands in fallow will constitute a contiguous area over which the animals may graze. However, this would not be necessary if rotating husbandry was not desirable (we shall return to this

---

[30] Thus, the actual number of fields thought of as being sown with the same crops could be greater than the recorded number of fields. G. Elliott, 'Field systems of northwest England', in Baker and Butlin, *Studies of Field Systems*, p. 43.

below). It appears, therefore, that the economic reasons for the practice of collecting the arable into large fields must be sought in two independent aspects of agricultural production : the practice of rotating husbandry, and the increasing returns to scale in grazing.

Even if it is possible to give an acceptable account for open fields in this manner, however, there is nothing so far that can explain the division of the arable into scattered strips. To attain large areas for grazing it is sufficient that the tenants gather their fallow plots into one contiguous holding. In the three field system this would be compatible with the farmer holding three roughly equal contiguous plots, one in each of the three fields. If everybody could agree on a crop rotation to suit them all, it would then be possible to achieve the proper grazing area. We know the obvious costs of scattering. Access to the plot is made more complicated, it is necessary to stake out the boundaries and leave a string of soil to mark them, the strips must be protected from transgressions and thefts, among other costs. Yet the farmers chose to preserve the scattering. No other explanation can be acceptable. To stay consistent with our postulate of wealth maximizing behavior, in our simplified static treatment of the problem, it is therefore necessary to show that the scattering itself provides some benefit that would not be attainable if the farmers held three consolidated plots, one in every field. Only in this way can we show that it is possible that the obvious cost of scattering may be more than compensated for by some less obvious benefit.

We have proceeded throughout on the assumption of private property rights in the arable fields, and have justified this by arguing that private property rights provide for a better incentive to the proper utilization of scarce resources than farming the arable with the help of hired labor. Whatever the reasons were, the direction of the evolutionary process is clearly that the villein, paying for the use of his land through labor services, not only establishes some rights in the land he has been allocated, but also comes to rent an increasing portion of the lord's demesne. It is the result of this long process that we have characterized, at the expense of some simplification, as private property rights in the arable. Given that the desired result is large scale grazing in the open fields themselves, the problem therefore becomes one of the combination of private property in the arable with communal grazing. This is very similar to the situation analyzed above with respect to the waste. The only difference is the further complication that the arable is divided into individually controlled lots. We have shown that one reason why collective rights were preferred in the commons was that, if the waste had been divided up into private holdings in severalty, it would have been more difficult to achieve large scale grazing. Each owner of land would be able to exert a strong bargaining position with respect to the other farmers and this would make the communal grazing costlier to attain.

This is identical to the problem encountered in the arable fields: large scale, communal grazing is desired within the context of land held in private ownership. Each owner must be persuaded to open up his land for grazing by others. By virtue of the increasing returns to land in grazing, this is the more important the more land the individual farmer controls, for the more he adds to the total grazing area. Yet the larger any one farmer, the greater his bargaining power *vis-à-vis* the rest. The more land he controls, the more he can cause economic damage to the others by threatening to pull out of the communal grazing.[31] This difficulty is inherent in the desire to combine private property with collective control. Its solution in the open field system, elegant in its simplicity, is scattering.

By requiring each tenant to divide his land into scattered plots, two important results are achieved: first, the benefits for any

---

[31] The benefits from large scale grazing can explain a feature of the open field system that puzzles McCloskey: 'If scattered plots had no advantages to set against their inefficiencies, however, and if there was no continuing mechanism to scatter them, even . . . piecemeal purchases and sales of land would produce in time consolidated holdings, for a peasant would seize every occasional opportunity to buy plots contiguous to his existing ones and to sell plots far from them or merely buy up land in one quarter of the open field. If he could not get exemption from communal rules of grazing and cropping, of course, consolidation might not be worth the considerable effort necessary to achieve it. Sometimes he could not, the difficulty being that the other commoners felt that his enclosure reduced the grazing left for their animals. It is unclear why this was so, for, as was noted earlier, when each person's stint was proportional to the amount of land he held in the open fields, withdrawal of some of it would have no effect on the remaining land available per animal.' D. N. McCloskey, 'The persistence of English open fields', in W. N. Parker and E. L. Jones (eds.), *European Peasants and their Markets*, Princeton, 1975, pp. 100–1.

tenant to withdraw from the collective, large scale grazing are radically reduced; secondly, the costs of organizing separate grazing on the individual plots become significantly greater. The benefits decrease because, by virtue of the increasing returns to grazing, there will be a greater reduction in output of animals for the farmer who decides to break away from the others if his plots are scattered about than if he held his land in one contiguous plot. The costs are increased because it is costlier to fence in small, scattered plots rather than one large piece of land of the same area. Scattering makes the costs of individual control over the holdings of any one farmer high because he has to fence off all the little strips from his neighbors in order to keep them out. What the scattering thereby achieves is the creation of an incentive for the farmer to participate in the collective decisionmaking and control necessary to regulate the use of the large grazing areas in both the commons and in the arable fields.[32]

So far we have identified two separate aspects of the scattering which facilitate the organization of an institutional arrangement to control the communal grazing : the decrease in hold-out power of any one farmer as against the collective that the scattering achieves, and the decrease in the net benefit of separating himself from the collective any one farmer would achieve. However, there is another interesting aspect of the scattering that would tend to work in the same direction. The most obvious costs of scattering are those associated with the changes in the physical aspects of the land consequent upon the division of one plot of land into scattered strips. Yet it is not clear that these costs were significant : the cost of scattering in the form of the reduction of the amount of crops to be had from the arable was probably insignificant. Perhaps the real cost of scattering lies in the neighborhood effects it creates as between the farmers. If one farmer does not do the weeding properly, his neighbor suffers. If one does not do his

---

[32] Curtler seems on the verge of realizing this when he points out the interesting idea that scattering constituted a way for the tenants to protect themselves against enclosure by the lord. This becomes explicable with the postulate of increasing returns to scale in livestock, for the tenantry would stand to lose much if the greater landowner withdrew : by ensuring that the lord's lands were intermingled with the others', the tenants could protect their grazing benefits. See W. H. R. Curtler, *The Enclosure and Redistribution of our Land*, Oxford, 1920, p. 120.

part of the fencing, all the others suffer. And there is the problem of theft of the crops and of moving boundary stones. Scattering inevitably creates such externalities, increases the causes for friction between the tenants and forces them to interact with each other at much closer quarters. For the purposes of decreasing the costs of organizing communal grazing, however, this only acts as an additional incentive, for the tenant who has a grievance against his neighbor will need some authority from whom he can demand redress. The community which organizes the control of the communal grazing is the natural body to which to turn for this purpose : the costs of scattering, that is to say, serve to increase the incentive to participate in the communal organization. The side effects of scattering are, as it were, imposed on the tenant in order to strengthen the viability of the collective decisionmaking organization.

A further argument in the same vein relies on the team work aspects of the arable fields. In many of the activities related to the normal tilling of the arable fields, the farmers would find it very natural to cooperate. This was the case for ploughing, where the farmers would join in ploughing teams, lending each other the oxen, the plough and the labor to go with it. The same was true for all the activities connected with harvesting such as reaping, gleaning, etc. Also, when the arable fields were increased by assarting, all the farmers would participate in the clearing of new soil. Certain later agricultural techniques, such as the floating of water meadows, relied heavily on the communal aspect of the open field system. Again, the imposition of scattering on the individual farmer would in a very natural way induce him to partake in such activities. The same notion can also be seen at work in the way payment to hired laborers was organized. In most cases, the hired man would get part of his remuneration in the form of some common rights, such as the right to put some animals on the commons, perhaps the right to till a strip or two. For many people, this formed an important way into the village community. It is very natural that it should be so. The institutions of the open field system would be kept strong and viable by admitting even the laborers into the community. By giving the hired man a piece of the communal cake, he is also given the incentive to see that the benefits of that very cake are not squandered lightly.

Hence, scattering does impose some costs on the farmer, costs that on balance cannot have been very substantial.[33] For as long as the communal ties remain strong, the costs of policing the implementation of the correct decisions about resource allocation cannot be significant, and as long as the institutions are viable, negative interactions between the farmers can be kept at a minimum. Nevertheless, the gain from scattering, open fields, and communal grazing on the waste would be quite important in terms of increases in the net output of animal products. Yet since scattering certainly would be costly for any one tenant, there ought to be an incentive for him to attempt to consolidate. Let us perform the following hypothetical experiment to see how that tenant would try to avoid the costs of scattering.

Suppose a tenant consolidates his holding by successive purchases and exchanges of adjoining lands until he achieves one contiguous area. The first effect to recognize would be the gains to the consolidating tenant; naturally, he has avoided the costs of

---

[33] Few estimates of the costs of scattering exist. Perhaps the most elaborate attempt to calculate the loss of income due to scattering is due to D. N. McCloskey, 'English open fields as behavior towards risk', in P. Uselding (ed.), *Research in Economic History: An Annual Compilation*, vol. I, 1976. He believes that the decrease in annual income to a typical peasant was of the order of 10 percent (p. 125). He bases this estimate on a comparison of rents on consolidated and scattered farms, and on an estimate of the elasticity of output with respect to the number of plots. What is the relevance of this 'guesstimate' for the present analysis? Consider: '. . . scattering was the root cause of whatever loss occurred: scattering, with its attendant inefficiencies, implied common grazing, with more inefficiencies, and scattering and common grazing together implied communal cropping, with still more.' McCloskey, 'The persistence of English Common fields', p. 87. In his estimates of the income loss to the typical peasant, McCloskey adds three components: costs due to (i) scattering, (ii) communal grazing, (iii) collective decisionmaking. However, if the ideas presented in this work are correct, it would not be right to add these costs in the way McCloskey asserts that they should be added, for element (ii), communal grazing, really represents a *benefit* since it enables the community to realize the increased income from large scale grazing, and element (ii), collective decisionmaking, is no cost at all, since the opportunity cost is zero: for we have shown in the preceding that even if the grazing areas were owned privately and the large scale grazing were to be attained, there would have to be some sort of collective decisionmaking to control grazing and crop rotations. So our argument would be that the net benefits from large scale grazing *offset* the costs of scattering, so that a greater net income is attained with scattering, and its attendant savings on bargaining and policing costs, than without scattering – quite contrary to McCloskey's estimates.

the scattering. What are the costs to the rest of the tenants of any one of them being allowed to consolidate in this fashion? Clearly, as long as nothing else changes, there would be no costs to the other tenants. The tenant who consolidated, however, would be better off not only because he has avoided the costs of scattering, but also by virtue of the fact that he has increased his bargaining power with respect to the other tenants, at least as long as their holdings remained scattered.

The argument thus leads to the following hypothesis: each single owner would have an interest in consolidating the scattered strips without at the same time having the others do the same.[34] Our analysis is reminiscent of the free rider problem where each recipient of the benefit stream from a collective good sees it in his own interest to dodge payment for what he consumes, since that would not reduce the amount available of the good. However, if everyone does this, any conceivable alternative way of producing the collective good turns out to be more expensive. Therefore, both joint and individual wealth maximization dictates that no individual be allowed to consolidate his strips. The infringements upon the economic activities of the rest of the community cannot be tolerated.[35]

The point is that for the peasants to realize the economies of scale in livestock, while still owning arable land of a size that is less than the total area of grazing grounds, collective decision-making is desirable. The commons must be stinted, since otherwise collective ownership gives each individual farmer incentives to overgraze. The costs of fencing must be allocated among the participants. The costs of supervision of the herds must be divided up. Nor would the fact that the tenants were allowed to con-

---

[34] Slater shows how this could be done: '. . . the holder of lands in common fields or common meadows, who fenced his holding, or parts of it, thereby prevented the other holders from exercising their rights of pasturing their cattle upon the fenced portions, without giving up his recognized right to pasture cattle on his neighbors' holdings, very likely indeed turning out all the more cattle in the summer and autumn, because better supplied with feed. . . .' G. Slater, *The English Peasantry and the Enclosure of Common Fields*, New York, 1968, p. 156.

[35] The argument is *not* that collective 'needs' override individual desires. Rather, the tenant is seen to accept the scattering as a price to be paid in order to prevent others from consolidating their holdings, and scattering is therefore consistent with individual wealth maximization.

solidate invalidate this statement : if a big herd on a large area is desirable, then consolidation means that it is still desirable. But with consolidation, the costs of attaining the collective decision-making would be higher since each tenant would wield more autonomous power to make individual decisions. With the holdings scattered the tenant would have no incentive to make individual decisions since the size of each plot would be too small.

It is clear that even with consolidation he could be forced by rules and regulations to participate in the collective. However, rules and regulations are not upheld by themselves. They have to be enforced. Regulations are maintained to restrict behavior, to prevent people from behaving as their personal incentive structure would induce them to behave. A way to avoid such costs is to change the incentive structure, and that is precisely what the scattering of the strips achieves. There is simply no incentive left for the individual tenant to make separate decisions when his strips are scattered. He participates naturally in the collective, so it is not necessary to make up and police any special rules to the effect that he should. As long as the returns to scale in the arable are not appreciable, the costs of scattering will not be very great. However, the transaction costs that the scattering saves on may very well be appreciable – there is no necessity to cajole stubborn individualists to go along with the majority. Scattering constitutes the least costly way to ensure the collective decisionmaking necessary to realize the returns to scale in livestock. Scattering achieves a change in constraints and incentives, rather than creating negative rewards in terms of punishment for keeping out of the collective. Tawney expresses this view well :

In 1405 some customary tenants at Forncett are fined 2s 2d. because 'they have made enclosures of their lands within the manor against the custom of the manor, on account of which action the tenants of the manor are not able to have their common there.' In 1518 the court at Castle Combe presents that three tenants 'have sown the common fields and kept them several without the licence of the lord, when they ought to be common, to the common damage'. . . . Enclosure made by one tenant on the open arable fields created a disturbance which was immediate and obvious. Indeed, if his holdings lay in scattered strips, separated from each other by the strips of his neighbour's how could he enclose at all? He would at once come into collision with their demands that his holding should lie open for grazing purposes after harvest.[36]

Heretofore, it has been natural in the literature on the open field, scattered strips systems of production to treat the scattered strips and the open fields as if they were part of the same general set of problems, that is to say, as if they necessarily imply one another as part of a comprehensive solution to the organization of production in the agricultural sector. The treatment here implies no such assumption. Indeed, it implies that the scattering of the strips and the collecting of the strips into several large open fields constitute the solutions to two very different problems. For the size of the fields is determined principally by two factors: the technology of fertilization in the sense of knowledge of ways to keep the fields productive without being forced to let them lie in fallow, and the returns to scale in the production of livestock, which require fields of a large size. However, the existence of scattered strips has very little to do with agricultural technology at all. Rather, it is the cost of enforcing and reaching the agreements of using the privately owned open fields collectively that determines the scattering. For it is clear that if another, alternative way of ensuring the individual tenant's participation in the collective could be devised that proved less costly than scattering, then scattering could be avoided while the benefits of large scale grazing on open fields could simultaneously be preserved. Scattering does not at all imply open fields, nor the other way round: one is the result of transaction costs in decisionmaking; the other of conditions of productive technology.

Recapitulation and review

The argument of this chapter is fairly complex, and we shall therefore summarize the analysis briefly before evaluating its claims. There are two separate parts: one pertains to the open field system itself and the economics of its special institutions, the other to the theory of transaction costs and property rights. Our stylized description of the representative open field village attempts to capture the main features of the system as it existed

---

[36] Tawney, *The Agrarian Problem*, p. 161–2.

not only in England but probably also over vast parts of Europe for a long period of time. The theory seeks to explain its rather surprising stability over time, space, and culture. This task has required some extension of the theory of property rights and transaction costs, and, in particular, we have endeavored to show that collective ownership and control over a scarce productive resource can be entirely compatible with economic efficiency and individual wealth maximization.

The production problem of the open field villages was to achieve as much output as possible from given productive resources, and to organize the interaction between the farmers so as to detract as little as possible from the output. There are two important technological keys to this system of production : the practice of rotating or convertible husbandry, and the practice of large scale grazing. If either of these two aspects is missing, the open field system should not be observed. This appears to be perfectly consistent with historical records, as has been discussed.

The achievement of the best possible output from the arable and from the grazing areas must also be combined with the best ownership structure. This is the other important element in an understanding of the institutional arrangements in the open field system. The peculiar mixture of private and collective property rights and decisionmaking chosen in open field villages was designed to achieve the economies of large scale grazing, combine this with the intensive use of the arable soil which convertible husbandry attains, without simultaneously sacrificing the efficiency characteristic of private property rights. There were private property rights in both outputs, animals and grain, and private property in the land in the arable, i.e., the land could be bought, sold, rented, leased, bequeathed, and inherited. Land used for arable purposes was primarily controlled by private decisionmaking. However, in almost all respects of the land that had any bearing on the production of livestock, communal control was exceedingly strong. Perhaps the element in the organization of production which is alien to modern eyes is the peculiar fact that the same piece of land would revert from private property to collective, and back again, in a well defined and controlled cycle. For the arable field would have its designated owners, each controlling a number of strips, whenever it was used for cropping. Yet as soon as that same field is employed for grazing, the whole collective exercises rights of common over private property. The reasons for this must be sought in the nature of the costs attached to controlling a resource which has two alternative uses, pasture and arable, where the optimal scale on which to utilize the resource in the two activities is significantly different, and in the costs of defining, protecting, and policing property rights.

Collective property rights have the inherent problem, referred to repeatedly above, of giving incentives to over-use. Unless some users are excluded, over-use will occur to the point where no rent accrues to the resource at all. It is therefore necessary not only to exclude users from the resources, but also to police the behavior of those who do acquire the right of usership. It is in this regard that the village council, or the village court, fulfilled an important economic function, the viability of which had to be preserved by the rules and regulations shared by the village members. There were a number of functions in the village community to be brought under collective control in order to guarantee efficient use of resources.

At a minimum, the following aspects relating to production in the village must be brought under collective control. First, in order to prevent the farmers from overgrazing the commons, as they would be wont to do with communal grazing, it would be necessary for the village council to limit the number of animals that each farmer could be allowed to put out on the common grazing areas. Usually, the simplest rule was adhered to whereby there was a direct proportional relationship between the number of animals any one tenant could put out in the village herd and the share of the arable belonging to him. This squares well with our postulate of mixed husbandry as being very natural for the agricultural techniques available during the period of the open field system.

Secondly, the council must see to it that the village members participate in the upkeep of the communal property. This would entail dividing up the costs of putting up fences around the grazing areas, the payment to the herdsmen, repairs on the commons, access ways, and the like. Again, the simple rule would be that the payment stands in proportion to the holdings in the arable.

Thirdly, it would be necessary for the village council to exercise some control over the exchange of strips in the arable. We have shown that the incentive would be for any one farmer to consolidate his strips, hopefully without the others doing the same thing. Yet if one has the incentive to consolidate, so does everyone. In order to keep the farmers from letting the communal organization break down, the community must exercise some control over such attempts to consolidate. It is clear that this was the case as permissions to obtain strips were secured from the court, and a fine paid for this. The implication is that the right to exchange freely was attenuated – only if certain conditions were fulfilled could such exchanges occur.[37] The council could agree to some consolidation, yet there is no apparent example of tenants who actually achieved full scale consolidation in this manner without the simultaneous enclosure of the whole village. Limited consolidation did occur in the arable, yet communal consent for such consolidation was required; for without such control it would be difficult to keep the communal organization viable.

Fourth, the village decisionmaking body must exercise some control over crop rotations. This is necessary to ensure that some large contiguous area will be in fallow each year; and that ploughing dates and harvest dates are reasonably well coordinated between the farmers so that rights of common of shack can be exercised. Limited enclosures in the arable, whereby certain tenants are allowed to deviate (more or less temporarily) from the generally accepted rotations may be allowed, but the principle must be that such deviations from the routine were exceptional and require special permission. There is much evidence that such was the case.

---

[37] Thus the theory here can explain an important feature of the system that puzzles writers like McCloskey: '. . . the lord of the manor had the right, except by local custom in Kent, to permit or to prevent the exchange of lands among those who held lands of him. It is not immediately obvious why this would have slowed enclosure, especially when the lord–vassal relationship acquired the character of a lord–tenant relationship, as it did increasingly in modern times: after all the landlord stood to gain, or at least did not stand to lose, from any increased efficiency of his tenants.' D. N. McCloskey, 'The economics of enclosure: a market analysis', Parker and Jones, *European Peasants*, p. 129. It can be noted that the lord would stand to lose from such consolidation if everybody loses the benefits from large-scale grazing.

Fifth, the village council must prevent the behavior of each farmer from having negative side effects in the arable fields on his neighbors. The village council must put fines on anyone not weeding his property correctly, or on anyone who cheats on any of the rules regarding the collective use of the property. Again, there is ample evidence that this was done.

Sixth, the council must appoint the officers of the village whose duty it will be to see to it that the transgressions are kept at a minimum. The reeve, the beadle and the shepherd were all hired by the village as a whole.

It should be noted that these actions by the council can, in principle, be predicted from the theoretical constructions above by virtue of the nature of the transaction costs we have attempted to specify. Each of the actions undertaken by the village council is designed to cope with some complications inherent in the nature of collective use of private property or of collective property. This becomes explicable within our theory as an integral part of a totality of relationship we have been able to specify.

In addition, we can now see the justification for the implicit voting rule : that the greater the share of the arable, the greater the influence of a particular owner. This is a voting rule that makes much sense in an economic perspective, for it best approximates that rule which is a guiding principle for establishing an allocation of resources which is also a Pareto optimum. In the standard setting of a Walrasian general equilibrium system, the dominant paradigm of modern economics, each economic agent 'votes' with his dollars for the scarce resources, and the theory shows how this leads to a situation that cannot be improved without changing initial wealth distribution.[38] It is also clear that if the commons were split up and divided into private property, then this is the voting rule that effectively would prevail, as it did after enclosure. There is a sound economic argument for this rule, for it places the greater influence in the hands of those that have most at stake, and who therefore have the greatest incentives to avoid inefficiencies. A democratic voting rule, one man one vote, is less desirable in this context, but may of course have important

---

[38] For a discussion of the implications of different voting rules, see J. Buchanan and G. Tullock, *The Calculus of Consent*, East Lansing, Michigan, 1962, *passim*.

optimality properties in a context where efficiency of resource allocation is not at issue, or is a goal of lesser importance. However, the very fact that democratic voting was not common in the open field villages is powerful evidence of the striving towards efficient use of scarce resources, or profit maximization in modern jargon, that must have been one of the overriding goals of the open field villages. Our theoretical framework can thus provide some explanation for the voting rule. It is a compromise between the difficulties of communal controls, with its poor incentives, and the potential bargaining problems of private ownership, with its better incentive and control mechanisms.

The most important claim for the theory, however, is that it identifies those benefits of the open field system that more than balanced the well-known, obvious costs of the scattering. Yet to make this claim good, it is necessary to establish the following two propositions. It must be shown that there would be no way of retaining the benefits of large herds on large areas without incurring the costs of communal decisionmaking and ownership. And it must be established that scattering constituted the least costly way of combining the private property rights in the arable with large scale grazing in the open fields.

Before the first proposition can be accepted, it is necessary to show that one single farmer could not operate at the same profit as the collective of the small individual farmers could. If it can be shown that mixed husbandry is profitable, and that one monopolist farmer would operate at higher total costs than the greater number of smaller ones, including the costs of managing the interactions between the latter ones, then the first proposition is established. It is perhaps easy to accept that mixed farming was profitable. If the same resources can be used alternatively in the production of both, the opportunity cost of producing the other, given that one is already being produced, will be very low. This was indeed the nature of the farming methods known in medieval England. However fertile the soil, some animals can be kept at very little cost in terms of decrease of output from the arable, especially since they provide carting and manuring services; but however rough and inaccessible the soil, so that it is mostly suited for pasture, there will always be some plots used for arable. Since the animals do not require constant attention, the opportunity cost in terms of land and labor in attaining some arable is small. There can be no doubt that mixed farming was profitable. It was universally practiced and not only in the open field villages. But if there were substantial returns to scale in grazing, why did not one grazier establish himself? Surely he could have attained the benefits of large scale grazing while avoiding the costs of policing and controlling communal interaction if he rented out the rights to the arable?

This would certainly have been the case if there was no inter-relationship between the two outputs. And here is the key to why mixed husbandry is profitable: grain and livestock do require each other as inputs, and can share other resources, land and labor, since neither, by the very nature of seasonal variations as they affect utilization of resources, is able to give full employment to labor or use the land continuously over the whole year. Consequently, if a monopolist grazier establishes himself in the waste and in the arable, and then proceeds to rent the arable to small farmers, the farmers can supply labor and land for use in grazing at low marginal cost. In order for the grazier to make use of this, additional transactions have to be organized between the farmer and the grazier. If such transactions were engaged upon, there would be further transaction costs, such as those of policing the grazing of the grazier's animals on the stubble in the arable or supervising the work of the hired farmers in making fences on the pastures. If the farmers were to engage in mixed farming on their own, such policing and supervision costs can be eliminated.

But if we are examining a system of individual farmers who collectively own the grazing areas, it is necessary for them to incur the costs of supervising and policing each other. In a community as closely knit as the open field village, this can never have been a substantial cost. The villages were relatively small, and the degree of personal knowledge between the farmers must have been quite high. It would be difficult, except for a very short time, for anyone to dodge participation in communal undertakings or to impose costs on the others by overgrazing. The transaction costs involved in the communal ownership and control cannot have been very significant as long as the village council acted forcefully. But a monopolist grazier would always have to accept the costs of supervision of hired laborers.

Only in this way, by attempting to specify the relevant transaction costs in each case, can we account for the fact that there were many farmers involved in mixed farming rather than just one specialized grazier. It is in the interplay of conditions of production with the costs of exchange and of communal interaction that the arguments behind the first proposition must be sought.

In principle, this also holds for the second proposition : that the scattering constitutes the cheapest method known for combining the private ownership of the arable with the collective use of it for grazing. We have argued here that the scattering achieves this by decreasing the incentives for private decisionmaking. Scattering makes it less desirable for the farmer to break off on his own, by raising the costs and by decreasing the benefits for him to do so. What the scattering thus achieves is to change the incentives. Conceptually, the frame of comparison is whether the same result can be obtained in a different manner. The age-old alternative to the carrot is the stick. Conceivably, the farmer could be compelled to obey the rules and regulations of the open field village community, so that the costs of scattering could be eliminated and consolidation achieved, and the benefits of large area grazing still realized. Yet such an alternative would create other kinds of transactions costs since it would be necessary to set up and police such alternative rules. Perhaps a strong lord of the manor could make it a condition of tenure that his tenants obey the rules and regulations as the lord would determine them. And if any lord ever tried, he would find out that his rules and regulations were unenforceable through the erosion of their foundations by the creation of rights at the common law. For by using the same piece of land for a long period, the peasant family would establish rights to the usership of that land, and the rights of the lord would become attenuated until the only right he could retain would be the one of collecting a rent on the land. Not even as against his own tenants could the lord ever uphold the rules and regulations that he would have to institute to keep a collective of large individual holdings functionable. With respect to freeholders it would be impossible for him to do so. There was no legal sanction to which the lord, or the court, could resort in order to force a recalcitrant freeholder with a large, contiguous piece of property to obey rules of crop rotation, fallowing, stinting, and so on. And even if there were such sanctions, they would be costly to enforce, and costly to finance from the resources of the other members of the community. But scattering bypasses all these enforcement problems in the simplest manner possible by creating a strong incentive for the preservation of the rules by the very owners of the arable themselves. What solution could be more elegant than this? As long as the open field village courts retained their authority, it appears certain that the other alternatives open to medieval peasants seemed more cumbersome.

The key to the choice of institutions lies in the way that they enable us to save on transaction costs. Fundamentally, this is the economic rationale for their existence. We have attempted to show how this was done in the context of the organization of production in the open field villages. Through the village council, the costs of decision about the use of resources can be kept low by the implementation of a known and stable rule for the reaching of collective decisions, and this saves on the costs of having a chain of private transactions carried out. The collective ownership of the waste saves on the costs of establishing joint usership of the grazing areas as compared with private property in the waste because it eliminates the necessity of a series of transactions. The scattering, while imposing some costs on the individual, saves on other costs by virtue of the fact that it creates the incentive for the farmer to keep the collective decisionmaking organizations strong and viable; thus, scattering saves on the costs of market exchanges required to make the farmer participate in the communal grazing. Whatever the force of these particular arguments, there can be no doubt that such a line of reasoning must be resorted to if an understanding of the choice of contractual arrangements in the open field village is ever to be gained. It is necessary to show the exact relationship between productive technology on one hand, and transaction costs on the other. Institutions serve to decrease transaction costs; that is well known. But the transaction costs themselves are related to the nature of the production processes under consideration. Both must be accounted for.

This accords very closely with the fundamental theoretical arguments in the property rights literature. The contention there is that the choice of property rights can be of importance only if there are transaction costs, for otherwise property rights of whatever kind will only affect income distribution, not resource allocation. The nature of transaction costs must be specified in detail before we can account for the choice of property rights in any particular context. This is the task we have attempted with respect to the open field system.

There remains the question of the degree of scattering : how many strips would each farmer hold? We have not so far dealt with this problem in any depth. What is important for the ideas presented here is that the size of each strip is so small as to provide little or insignificant grazing in itself. Naturally, the number of strips that a typical peasant would hold would then depend on his wealth as the primary determinant.

The most intensive study of the degree of scattering in recent years is the work of McCloskey, referred to above.[39] He claims to have shown that the six or so strips that a 'typical' peasant would hold in each of the three fields in a three-field village is consistent with risk aversion as the 'cause' of scattering. Two points about McCloskey's claims are relevant in this context. First, contrary to what is apparently implied by some of the writings of McCloskey, all econometric analysis can ever do is to show correlation. There is absolutely nothing about causation that can be inferred from the estimates. It follows that even if scattering is *consistent* with risk aversion, it cannot be inferred that risk aversion is also the *cause* of scattering. Secondly, McCloskey's estimates are not contradictory to the ideas brought forward here, and cannot therefore be used as a counterproposition that 'falsifies' a theory of scattering which maintains that it constituted the least costly way of dealing with certain bargaining and decision costs resulting from communal interaction. For if the community decides on scattering as an intelligent course of action, then it would naturally be prudent to minimize the costs that scattering implies. A natural way of doing this would be to lay out the strips so that a minimization in the variation of income is achieved, or, in other words, so that risk is minimized. McCloskey's estimates are therefore quite consistent with the ideas presented here, for all they show

[39] McCloskey, 'English open fields as behavior towards risk.' The estimate of six plots per typical tenant occurs on p. 157; the assertion of causation on p. 165.

is that the degree of scattering is determined so that risk is minimized. What has been presented in this chapter is a theory for
the preservation of scattering that uses transaction costs as the
basic element. McCloskey's cliometrics can therefore easily be
submerged into this theory as an integral part. The converse does
not follow, however. If the preceding ideas are accepted, then
McCloskey's claim to having discovered the 'cause' of scattering
is incorrect – i.e., the ideas here cannot be made a subset of
McCloskey's work. But his results on scattering could indeed be
made an integral part of our theory.

We may also relate the arguments of this chapter to the traditional Marxian interpretation of the open field system as a vehicle
for exploitation and income redistribution. At the beginning of
this chapter, it was pointed out that the reasoning here only requires the comparison of two known alternatives, the open field
system and enclosed farms, and that the contention is that the
open field system, given the conditions specified here, was efficient
relative to enclosed farms. The argument is not that the open
field system was the most efficient possible system of agriculture :
that is an issue that cannot ever be settled. It follows that we have
not logically ruled out the possibility that the open field system
was inefficient from an overall allocative standpoint, i.e., inefficient relative to some hypothetical global optimum, but efficient
only from an income redistribution point of view, as determined
by the lords. That issue cannot be addressed in a meaningful
manner, since the total choice set of agricultural institutions in
which the open field system is but one element cannot be properly
defined. Hence, it follows that there is nothing *per se* in the transaction costs and property rights approach adopted here that is
inconsistent with either the traditional Marxian interpretation of
the open field system or with a view in the North and Thomas
vein that would also look at the system as efficient only from an
income redistribution aspect. If it can be shown that a more
efficient alternative to the open field system indeed existed, then
the arguments presented here will still retain whatever force they
presently have, for it will still be true that the open field system
was efficient relative to enclosed farms, given the conditions
specified here. That means that even the Marxian view as well
as the more modern formulation of similar themes by North and

Thomas could, in principle, be submerged into the present analysis as an integral element, should it ever be shown that a more efficient system of agriculture was indeed available to open field villages and enclosed farms alike.

However, it should be noted that the theory presented here really vests some of the crucial features of the open field system with important income distribution functions. Ultimately, the reason why the commons were used communally and controlled collectively was that this communal ownership served to protect the rights of the small and the weak as opposed to the rich and the powerful, for it prevented the large owner from using his better bargaining position to appropriate some of the gain that rightfully belonged to a small owner. Our interpretation of the efficiency of scattering also rests on the ability of scattered strips to limit the bargaining power of the larger owners. Thus, rather than being instruments by which the powerful gained from the weak, they were tools for guaranteeing that the gains to the small owners remained in their hands, i.e., for ensuring that the income redistributions stressed by Marxian analysis as well as by North and Thomas do not occur. However, this does not exclude the possibility that, relative to some other alternative, the open field system contained additionally some exploitative features.

The origins of the open field system

Before discussing the course of enclosure, we shall address the question of the origins of the system. Again, it must be stressed that no definite results will ever be achieved in this area, for the origins of the open field system are buried in antiquity, and no records of its introduction have been found. Here the account will necessarily be speculative, although this is a defect shared with all other accounts of the origins of the system. Even so, it may have some value, for some interesting differences as compared with the standard accounts will emerge.

If we accept Joan Thirsk's strict definition of a common field system,[40] it is necessary to account for the following four distinct

---

[40] Joan Thirsk, 'The common fields', p. 1.

elements : strips in the arable, common of shack, common waste, and communal regulation. The traditional explanation then presumes that the communal ownership of the soil is the oldest element of the four, possibly since communal ownership was a hereditary feature of the Germanic tribes that originally laid out the open field system, and that from the communal ownership of the waste follows the collective grazing. A distinct feature is the introduction of the scattering. Here, one *ad hoc* explanation or another is made to account for the occurrence of scattering – of the type referred to above (partible inheritance, assarting, common ploughing, desire for equity). Once scattering has occurred and grazing becomes more scarce as the population increases, common grazing of the arable becomes necessary; and this in turn requires communal controls.

The account consistent with our theory would be somewhat different. Instead of relying on some legendary features of our culture, it is quite sufficient to argue that communal grazing of the waste is a natural element of mixed farming, for reasons discussed above : the net output from the grazing animals is probably greater since fencing costs and supervision costs show some elements of returns to scale, and collective property proves cheaper than private in reaching the best solution to a large grazing area for a large herd of animals, for reasons inherent in the costs of decisionmaking. It is likely, therefore, that communal grazing of the waste is the oldest element of the open field system, not for cultural reasons, but for strict economic ones : this is the cheapest method of obtaining the best output from animal husbandry when combined with arable cultivation.

Communal rights to a scarce resource create the usual problems of over-use and underinvestment, and it is quite likely, therefore, that some sort of communal control over the waste is the second oldest element in the evolution of the open field system. Even if the commons were not stinted, as no doubt they were not in the beginning of the system, it would still be necessary to share in the costs of supervising the animals, in making fences, gates and roads. It would also be necessary for the village community to act as a unit in relation to other villages in disputes about as yet weakly delimited rights of common in certain areas. A communal decisionmaking organization would therefore be the next natural step. Communal ownership requires collective interaction and decisionmaking.

The next element is probably the introduction of grazing in the arable : common of shack is a primitive form of alternating husbandry. As was pointed out above, it is not costless to introduce this type of regime, in spite of the fact that the animals add fertilizers to the soil, for it would be necessary to fence the holdings in the arable, to provide access roads and the like. As long as there is plentiful grazing to be had on the waste, the costs of rotating husbandry can be avoided. Hence common of shack is probably a later development than common grazing on the waste. Again, common of shack is also founded on the wish to have a large area for the animals to graze on, for otherwise each farmer would be quite content with grazing only his own land, and would not stand to gain much from having the right to graze his neighbor's property. The formation of large open fields with grazing on the fallow can therefore be expected to have followed population growth quite closely. The conclusion, then, is that common of shack is introduced only when the grazing on the communal waste lands becomes scarce. In this situation the intensive utilization of the soil introduced by rotating husbandry in the arable, i.e., common of shack, is a natural consequence.

Having introduced common of shack, it would then also be necessary to impose some rudimentary controls on the cropping arrangements in the arable. If the best grazing area is large, it would be natural to form large open fields, and the imposition of a requirement to divide his holdings into two parts, one in each field, was perhaps the first control the medieval farmer was subjected to. Once the two field system had been formed, the next step would be to safeguard the rights of grazing in the arable, and this would require some regulations about cropping; at a minimum, a decision must be made as to what field will be fallow, and when the cropped field will be opened for grazing after harvest.

Therefore, it would seem that the scattering of the strips is the last of the elements to be introduced into the fully-fledged open field system. There is no reason to assume that scattering was already present in the arable when common of shack became desirable : the situation portrayed is quite consistent with initially consolidated holdings. However, as soon as common of shack is introduced there will be the additional problem of ensuring that all farmers participate in the collective grazing, and the collective decisionmaking and control over cropping arrangements. Exactly how scattering was introduced will never be known; but the account given here implies that scattering was imposed on the villagers so as to make the collective decisionmaking institutions viable and effective. Exactly how this was done is perhaps not so important. For example, since we have shown that scattering had powerful reasons for *persisting*, almost any initial reason yielding scattering is quite consistent with the theory. That is to say, partible inheritance, assarting, common ploughing, or a desire of equity may all be partial answers to the quest for the origin of scattering that can be used in conjunction with the present arguments for its persistence. On the other hand, our theory would imply that scattering is *imposed* on the owner of the strips, since he would be better off if he could consolidate without anyone else doing the same. It is interesting, therefore, that all traditional explanations for scattering imply some element of forcible imposition of scattering on the peasant – some custom or tradition is invoked to explain why scattering occurs. There is no particular reason why it should be presumed that scattering was introduced *suddenly*. On the contrary, it would be very natural to assume that scattering grew out of partible inheritance, for example, and was found to be a convenient way of preserving the organizational arrangements of the open field system, and so was shown viable.

There is one important piece of evidence that is consistent with this account. Several writers[41] point out that there was at quite late dates what appears to be a formal restructuring of the open field villages, where land seems to have been shifted around, fields rearranged, strips exchanged, etc. In these examples one often finds that one owner has the same neighbors at different locations in the same field. This would seem to point to a conscious effort to protect the function of scattering, as it was maintained in the reorganization of the village. Naturally, scattering may very well

---

[41] Baker and Butlin, (eds.), *Studies of Field Systems*, Cambridge, 1973, note this in their postscript, pp. 651–3. They refer to several of the writers in the volume who have observed such reorganization in various places throughout England and Wales.

have existed before then, as for example Hoffman points out,[42] but it is not until the village entered the mature phase of the gradual introduction of the open field system – the phase that includes all the stylized facts referred to in the preceding – that scattering would come to play its crucial role as an element which kept the social fabric cohesive.

[42] R. C. Hoffman, 'Medieval origins of the common fields', in Parker and Jones, *European Peasants*, p. 43.

5

THE ECONOMICS OF ENCLOSURE

In the previous chapter we presented a theory of the scattered strips and the open fields. Throughout, we have stressed the fact that any theory of a particular institutional organization that attempts to construct an explanation in terms of economic efficiency cannot be considered satisfactory unless it is also consistent with those factors that made that organization undesirable, so that it was eventually replaced by another one. With respect to the open fields there is a remarkably clear-cut series of events that marked the end of the system – the enclosure of open fields. Clearly, the previous attempt at a theoretical construction will amount to nothing but a tedious game, unless it can be shown that this particular approach is also consistent with the known facts of the enclosure movements.

Our explanation of enclosures may therefore also be regarded as an attempt at an empirical test of our hypothesis about the nature of the open field system. As such, it will be no more than an initial step; we do not attempt a full-scale statistical investigation. The aim will be set at a comparatively low level, as judged against the standards of the modern cliometric school. It will suffice to resort to another set of 'stylized facts', in this case those features of the enclosure movements that are accorded a fairly broad consensus among the specialized historians of the primary source material and the local history of enclosure.

However, one of the more awkward problems is that there really is no unique definition of enclosure that is employed by all writers. The concept is not as clear-cut as could be desired. For example, there is the problem of how to treat closes in the open fields. Are they to be regarded as a breakdown of the basics of the open field system, or as a method of making that very system of production work more smoothly? There is a further problem with assarts from the communal waste. Are they to be treated as a way of enclosing communal property into private, or as the result of a collectively made decision to expand the arable at the expense of the waste? Unless we can arrive at a conceptually clear definition of enclosure we shall have to cope with ambiguities such as these.

For our purposes, therefore, we shall define enclosure to mean the simultaneous occurrence of the following two events: the consolidation of scattered strips, and the abolition of communal rights and decisionmaking. Collective rights could not be abolished unless the owner of such rights voluntarily relinquished them;[1] this was not possible unless there was a simultaneous abolition of communal grazing rights in the whole village. We shall therefore understand by enclosure the enclosure of the whole village, and not a piecemeal, temporary permission to deviate from communal rules of crop rotations and grazing stints on parts of the open fields. Since such irregular practices required the explicit consent of the community to be practicable, the inference will be that they were rather an example of the strength and viability of the open field system than of its gradual breakdown.[2] Consequently, temporary enclosures and permanent assarts must be regarded as a natural offshoot of the open field system,[3] and one that in no

[1] D. N. McCloskey, 'The economics of enclosure: a market analysis', W. N. Parker and E. L. Jones (eds.), *European Peasants and their Markets*, Princeton, 1975, p. 130, says: 'All those who owned rights of any sort in the open fields had to be brought into the agreement for it to be a legally binding contract, for the law quite reasonably required that a man's consent be obtained before the community could meddle with his property.'

[2] In essence, this issue turns on the influence that the community exerted over limited enclosures in the open fields. Yelling says: '. . . piecemeal enclosures covers a range of processes in which action is taken either by individuals or by groups of varying sizes up to, but not including, the whole body of proprietors. (. . .) The vital feature of enclosure was that it meant withdrawal from *common* husbandry practices, and if this step were not taken in common then this is most significant.' J. A. Yelling, *Common Field and Enclosure in England, 1450–1850*, London, 1977, pp. 6–7. However, later on Yelling makes the point that even in these cases the community could not be entirely disregarded: 'Piecemeal enclosure, too, always requires some degree of consent, or at least acquiescence, on the part of the whole body of landowners, but this is only permissive.' (p. 80).

[3] Tate says: 'In later years and in some manors it was on an occasion alleged that the custom of the manor permitted enclosure on the part of anyone, lord or tenant, provided he surrendered common rights upon the land still remaining open, and regularized the whole process in advance by obtaining

significant way implies enclosure in the full sense of the word.[4] We shall understand by enclosure the dissolution of the village community in favor of a system of private individual producers, with decisionmaking rights limited only by common and statute law, but not by village bylaws and rights of common originating in the interaction of the farmers themselves.

The ensuing procedure will therefore be to subject the theory to certain disturbances, and to see if the comparative static method predicts enclosure as the wealth maximizing alternative under the new circumstances. This method should therefore result in the identification of a well defined set of exogenous disturbances that, in the context of the theory, is consistent with enclosure. If that is the case, it will then be possible to derive some testable empirical implications: did the major waves of enclosure movements coincide with changes of that nature? If not, the theory must be rejected. Conceptually, at least, it is possible to specify such occurrences as will be inconsistent with the transaction costs structure presented here, in spite of the somewhat awkward fact that empirical estimates of transaction costs cannot be had. The transaction costs assumptions will therefore have to be tested in a more roundabout way.

The role of technology

Our theory rests on two types of 'technological' conditions. One is the technology of production, the interrelationship between the outputs, livestock and grain, the other is the reason for the quota-

---

the sanction of the court.' W. E. Tate, *The English Village Community and the Enclosure Movements*, London, 1967, p. 36. There are two significant things to note about this: first, if such enclosures could be made and the open field village still remained viable, the implication is that in general it was not worth it to relinquish common rights in the neighbors' lands; secondly, enclosures could not be obtained without the permission of the court, so the community kept its control even over closes.

[4] E. C. K. Gonner would support this interpretation: 'It does not follow, however, that land even when so inclosed would escape all incidents of common; indeed, the contrary must have been true in many cases, as where shack beasts being turned on to the arable after harvest had a right of entry into the inclosed plots through the bars or by the gates.' *Common Land and Inclosure*, London, 1966, p. 38.

tion marks, for it relates to the methods of reducing the transaction costs resulting from combining these methods of production
with different types of ownership and decisionmaking. Our first
task would seem to be to investigate these technologically determined features, and to ask what technological changes might
make the open field system give way to a different mode of production.

Several such technological changes are conceivable, albeit not
all of them plausible. For example, anything that would make
the optimal size of the grazing area equal to the size of the arable
would make scattering and collective ownership unnecessary. If
such a change occurred, say by suddenly introducing increasing
returns to scale in the tilling of the arable, the whole village ought
to be consolidated into one single economic enterprise. Scattering
and collective grazing would then disappear. Another such change
would be if rotating husbandry one day was no longer desirable,
for then there would be no grazing in the arable, and the scattering would consequently serve no purpose. In this case, however,
the communal grazing on the commons might be preserved, so
the court would be necessary to regulate this aspect, but cropping
rules, for example, could be done away with.

The problem with reasoning along these lines is that it leads
to the inevitable logical implication that enclosures ought to have
swept the English countryside relatively rapidly; for once the new
technology was introduced in some village, it ought to have been
copied in the next, and then the next, until the new method of
production was implemented everywhere over a comparatively
short period. It becomes impossible to explain why enclosure took
over four hundred years at least to accomplish. Furthermore, such
a change ought to have been implemented in some particular
area, and then spread outwards from this center; and it should
be possible, therefore, to identify roughly when and where such
a technological change occurred. Again, this is wholly inconsistent
with the gradual nature of enclosure, and its geographically dispersed occurrence.

These points are of a general nature with respect to certain
changes in conditions of technology of production. It has already
been noted how important the link between the production of
livestock and grain was in the open field system. Perhaps the

most important part of this link is the role that livestock played in the production of grain – in many areas the animals were for a long time the only cheap source of fertilizers that kept the productivity of the soil at a reasonable level. Although other fertilizers may have been available, there can be little doubt that in many places animals were the only source that allowed grain production to remain profitable. However, it is of course possible (at least in principle) to envisage a change of agricultural technology that made this link between livestock and grain of no particular consequence. For example, certain crop rotations may be discovered that preserve the nutrients of the soil in a better manner, or certain new crops may be introduced that perform this same function. There is little doubt that such considerations may have played an important role in the later history of enclosure, and we shall therefore return to them below. However, it must be noted here that a theory that makes such changes in technology the sole and exclusive cause of enclosure fails to account for the long time the open field villages took to submit to the new technology. If the new techniques indeed were so profitable as to make a change to a new system desirable, it is difficult to understand why the open field system so stubbornly resisted change and persisted in its old inefficient ways.

In the static framework within which the theory is couched, it becomes impossible to reconcile any radical technological change that altered the intimate relationship between the two major classes of ouputs, livestock and grain, with the very gradual growth and only occasional concentrated outbursts of enclosure. Any technological change that would have made undesirable the combination of collective and private property rights and decisionmaking, ought to have happened instantaneously. It is therefore not possible in the present theory to specify a single change of technology so that enclosure is the resulting wealth maximizing solution, while remaining consistent with the gradual nature of enclosure, and its seemingly random geographical dispersion. Furthermore, not only have we derived collective decisionmaking as consistent with private wealth maximization, but we have also observed that it provides for enough flexibility to allow individuals who wish to deviate from crop rotations, accepted ratios of arable to pasture and the like, to do so with the explicit permission of the court. Technological change, that is, is quite consistent with the preservation of the open field system.

In the context of the present theory, we must therefore unequivocally reject any hypothesis of enclosure that relies solely on technological change. The theory is quite consistent with the introduction of new methods of production into the system : any new technology that does not radically alter the relationship between the two outputs in terms of scale of production can readily be adapted within the existing institutional structure. The implication of this is that the villages remaining in the open field system ought to be seen to adopt the new techniques as they come along much at the same pace as the villages that enclosed. However, the open field system may, because of its peculiar institutions, be prone to a *particular* type of technological change – for example, to the adoption of techniques that have strong aspects of team work, or changes that would need to be implemented in the village as a whole to be economically viable. This raises the difficult question of the extent to which technological change is an *endogenous* variable, and the role that the design of the institutional framework, within which productive decisions are taken, plays in determining the course of technological change. This can explain why the newly enclosed villages were seen to adopt many new methods that were never implemented in the open field villages themselves. The implemenation of such new techniques cannot *per se* be taken to imply any slothfulness on the part of the decisionmakers in the open field system. Techniques that were not implemented in the open fields may well have been suited for a different type of organization of agriculture. The speeding up of 'technological change' which according to many observers accompanied enclosure can therefore perhaps be interpreted as the readjustment of agricultural technology to a new institutional framework.

It is also possible, in principle at least, to inquire into the effects of changing the methods for reducing the transaction and decision costs that played such an important part in our presentation. However, any attempt to conceive of alternative means of reducing transaction costs in this manner becomes a near-impossibility.[5]

---

[5] For a discussion of related topics, see H. Demsetz, 'Information and efficiency: another viewpoint', *Journal of Law and Economics*, vol. XII,

The proper question to ask is whether at any one time an alternative way of organizing production, given the production function for the two outputs and the amount of resources available, was conceived, so that an alternative set of institutions would be preferable to those of the open field system.

Again, if this was our theory of the enclosure of open fields, it would be necessary to go back to the drawing board and start again, for if there ever was such a change, the open field system would have tumbled in a very short time, for the very reasons we discussed above with respect to a change in the technology of production. Any such alternative way of organizing the institutions which guide resource allocation in the open field system should have swept the rest of English agriculture in a very short time. Certainly it would become totally inexplicable why certain areas of England seemed to be introducing the open field system at the same time that other areas were abolishing it by enclosure. So quite apart from the very contrived nature of the exercise of conceptualizing different methods of reducing transaction costs, it would be necessary to conjure up a fantasy of how such alternative modes of organizing institutions actually took hundreds of years to communicate and introduce. However, cases such as these should be considered, if for no other reason than to point out the important role that the transaction costs play in the theory, and to insist that the alternative means of coping with such costs present an economic problem on a par with the choice of techniques for the production of different outputs.

This brief discussion of the influences of the technological features of the theory has provided some additional insights into the requirements that an adequate theory of enclosures must meet. First of all, it must provide a gradually developing element of change; secondly, it must be of a nature that can explain the wide geographical dispersion of enclosures, as well as the important fact that some areas introduced the open field system at a time when it was already declining in other areas; thirdly, the change must be of a kind that upsets the balance in the production of the

---

1969. Demsetz would point out that we are committing the 'fallacy of the free lunch', 'the grass is always greener fallacy', or 'the people could be different fallacy' if we assume that a different type of behavior could reduce transaction costs.

two main classes of outputs considered here, so that the institutions designed to cope with the alternating use of the same resources in two different productive activities no longer prove necessary. Again, it is natural to look for a conceptually possible change in the constraints on production in the open fields.

The role of markets

A prominent feature of the theory is that both outputs are presumed to be produced in all villages, regardless of the particular quality of the soil in any particular village, although it is possible to allow for somewhat different mix of the outputs. The striking feature about this is that, although the soil qualities of England differ greatly between regions, the framework here does not predict any specialization in the production of one of the two outputs (an explicit assumption in the foregoing has been that *both* outputs are produced as *cash* crops). That is to say, not only do we assume that the open field villages produce enough of both outputs for their own consumption, but also enough for trading with surrounding areas, at least to some degree. The remarkable fact about the open field system appears to be the lack of specialization in the production and sale of one of the two outputs.

There were areas in England that specialized very early in the production of one of the outputs considered here, and in these areas it is extremely doubtful whether the open field system was ever in use at all. In Kent, for example, the arable fields were laid out on a pattern of rectangular fields, with no scattered strips, as opposed to the irregularly shaped open fields in other areas; so no conclusive proof can be brought to bear on the question whether Kent ever had the open field system.[6] Specialists do not seem to believe that it had. However, Kent served as the granary of London. Within easy reach of the city, this county had the largest single market in England, and it is no wonder that Kent was one of the first counties to specialize in the production of one

---

[6] A very good analysis of the Kentish system is given by A. R. H. Baker, 'Field systems of southeast England', in A. R. H. Baker and R. A. Butlin (eds.), *Studies of Field Systems in the British Isles*, Cambridge, 1973, pp. 337–430, *passim*.

F

of the two outputs.[7] The case is much the same in the regions
that specialized very early in the production of livestock : the hilly
uplands, with abundant pasture. These were the first to abolish
the open field system.[8]

The point appears to be general : whenever the open field system
was abolished, it was followed by some degree of specialization.[9]
The converse is also true. If a particular area of England was
found to specialize in the production of either grain or livestock,
it did not have the open field system. The fact that the open field
system thus seems to have been explicitly designed to cope with
a system of production of two interrelated but distinct outputs,
provides the clue to the question of what in the end made for both
its apparent stability and its rather dramatic disappearance. The
implication is that we should look for a correct hypothesis of
enclosure in the decisions about what to produce in the open fields.

We have discarded explanations exclusively based on technological change, and we cannot assume radical changes in the
endowments of means of production. There is really only one
element left : the extent of the market, and the influence of relative prices. If the present model is to be able to explain enclosure
at all, it must be shown that a change in market conditions
actually will yield consolidation of scattered holdings, and the
extinction of communal rights. As long as there are appreciable
differences in the composition of the soil between different regions
of England, and these regions produce both outputs as cash crops,
there will be comparative advantages in production that remain
unexploited. Potentially, each such region could increase its total

---

[7] See e.g., C. S. and C. S. Orwin, *The Open Fields,* Oxford, 1967, p. 64.

[8] See e.g., M. M. Postan, *The Medieval Economy and Society,* Berkeley and
Los Angeles, 1972, p. 52, where he says: 'Of the local variations none
diverged more radically from the prevailing system than the use of land as
closes, i.e., compact plots which their occupiers held and cultivated in
severalty, outside the framework of the common fields. In some areas of
England the closes could represent the prevailing form of landholding and
cultivation, as they often did in pastoral uplands. Such arable activities as
the inhabitants of these regions carried on alongside of their pastoral
pursuits were frequently conducted in enclosed crofts and in severalty.'

[9] On this point, see, e.g., R. B. Smith, *Land and Politics in the England of
Henry VIII,* Oxford, 1970, p. 19; E. Kerridge, 'Agriculture 1500–1973', in
R. Pugh (ed.), *A History of Wiltshire,* (Victoria History of the Counties of
England), Oxford, 1959, vol. II, p. 54; J. D. Chambers, *Nottinghamshire in
the 18th Century,* London, 1966, pp. 137–72.

income if it found a way to specialize in the product in which it has a comparative advantage. It remains to show that specialization also would yield enclosure, in the context of the model presented here.

Contemplate a village whose comparative advantage, if trade on a sufficient scale could be resorted to, lies in the production of grain, and let us suppose that there is an increase in the demand for both outputs. Given that the comparative advantage of this particular village lies in the production of grain, some other village within the same market area would have a comparative advantage in the production of livestock. When demand increases for both outputs, we should expect to see each village increase the production of that output in which it enjoys a comparative advantage. A further point to note is that the geographical extent of the market also increases; that is to say, a greater region is now brought into the confines of being influenced by the same market determinants, and there should be a net increase in the number of villages that must be considered when the comparative advantages of each village are decided. The anticipated result should be that the villages with a comparative advantage in the production of livestock should expand their production of livestock and contract the production of grain, as the villages with a comparative advantage in grain expand the production of grain and contract that of livestock. Each village, which produced both outputs for the market before the increase in the demand for the outputs, ought now to specialize in the production of one of the two, and let the other village, with the different production advantage, specialize in the production of that output. Indeed, with competition, it will be forced to do so.

An increase in the size of the market, that is to say, should yield specialization in production in each village, according to the comparative advantages inherent in its location and soil qualities. It is of course in the nature of agricultural production that the opportunity cost involved for even a specialized farmer to produce enough of the other output for his own consumption is extremely low, so it would be quite reasonable to expect that some of both outputs will always be produced in all villages. It would be to militate against the actual conditions of agricultural production opportunities to assume that specialization would be so complete as to actually eliminate the production of either of the two main outputs. What is considered here as specialization is rather concentration in production on one of the outputs *for sale* to a market which is supplied from other villages in the same market area with the other output. It is specialization in production for sale that is implied here rather than perfect textbook specialization.

In a village which thus specializes in the production of grain, the decision will be taken to decrease the pasture areas, and to expand the arable by sowing in the former grazing grounds. If the expansion of the market is of a magnitude which is not sufficient for commercial production of livestock to be totally abandoned, the only result that should be predicted is just this expansion of the arable at the cost of some pasture area – in other words, there would be no desire also to abolish the collective control and decisionmaking institutions as long as the production of livestock is still financially important. However, if the desired increase in the production of grain is so large that the only livestock that will be kept are the ones necessary for self-sufficiency, then the whole production problem faced by the village community is suddenly very different.

For no longer would it be necessary or desirable to alternate the same plot of land between the production of two outputs to be sold commercially. In the new situation, the one that is produced only for domestic consumption would be subjected to the conditions imposed by the desire to obtain the best output possible from the cash crop. In the case of specialization on grain, that is, it would no longer be desirable to incur the costs of scattering simply in order to achieve the large-scale grazing in the open fields, for the production of livestock suddenly takes on a role that is of secondary importance. Consolidation of scattered holdings is precisely what would be expected under these conditions. On the other hand, there is no longer the need for the communal grazing grounds in the commons. The commons should be divided between the farmers owning parts of the arable, and they should be adjoined with the arable fields. Hence our theory is entirely consistent with consolidation of scattered holdings and the abolition of communal property rights when a disturbance, exogenous to the village, which makes specialization desirable is specified.

In such a situation it can readily be seen that the collective decisionmaking institutions, the village council or the manorial court, would no longer serve any relevant function. The farmers would have little cause for interaction with each other after consolidation and parcelling out of the waste. Each farmer would be expected to conduct his affairs in severalty, and what little communal affairs are left after consolidation, division of the waste and specialization in production, can be handled in a more informal manner. For a change that increases the market size to such an extent as to make specialization economically desirable it would therefore seem that the present theory predicts all the important effects of enclosure : consolidation of scattered holdings, extinction of communal rights and decisionmaking, abolishment of the manorial court, and regional specialization in production.[10] However, there is one further interesting little point : the theory also predicts an increase in the average size of the farm producing grain.

In Chapter 4 the point was made that, for the purposes of making the collective decisionmaking body strong and viable in order to regulate the use of the communal property, it would be desirable to pay the farm worker, at least partly, in terms of ownership of the land of the village itself. Thereby the worker would get a stake in the village, and would have the incentive to use his productive resources as efficiently as possible.[11] There may be a cost involved in such a method of payment if the size of the holding belonging to a part-farmer, part-worker is smaller than

[10] It is a necessary condition, one that any acceptable theory of the open field system and the enclosure movements must abide by, that it can predict specialization in *either* output after enclosure : it has been shown that, over all the centuries of enclosing activities, enclosure was for pasture just about as often as for arable. Smith, *Land and Politics*, p. 19; Kerridge, 'Agriculture 1500–1793', in Pugh, *A History of Wiltshire*, vol. II, p. 54; Chambers, *Nottinghamshire in the 18th Century*, pp. 137–72.

[11] Slater states : '. . . in the open field village the entirely landless labourer was scarcely to be found. The division of holdings into numerous scattered pieces, many of which were of minute size, made it easier for a labourer to obtain what were in effect allotments in the common fields.' Slater, *The English Peasantry and the Enclosure of Common Fields*, New York, 1968, p. 130. Naturally, these small parcels did not constitute viable economic enterprises, and would be expected to disappear in the enclosure. In the open field system, however, they could well serve an important economic function in keeping the system viable.

the optimal unit of production. That cost can be avoided after enclosure, for when the collective decisionmaking institutions disappear, there would no longer be any reason for the payment to the worker to take such form as to give him an incentive to participate in the collective decisionmaking. Consequently, the land belonging to a small farmer or squatter ought to be bought by the larger ones.[12]

So far the discussion has only concerned a village that has comparative advantage in the production of grain. We have shown that the model is consistent with all the main elements of enclosure under the conditions of specialization in the production of grain. It remains to be seen whether this will be the result also for a village that finds it profitable to specialize in the production of livestock. The argument runs along exactly the same lines : when the market expands, a village with natural conditions such that it has a comparative advantage in the production of livestock should gradually expand the areas used for pasture, and then enclose when there is so little left of the arable that it is economically useless for cash crops. It is clear that in this case the ultimate result would accommodate only one or a few graziers in each village specializing in livestock, for in the open field village the grazing is done on a communal basis, and the scale of operation is the whole village. Enclosure would not alter this technological fact and only one grazier ought therefore to survive the abolishment of open field husbandry. Naturally, after only one grazier has established himself in the village, there will be no need for the scattered strips, the communal ownership of the waste, or the communal decisionmaking institutions designed to cope with the control of a collective resource. The conclusion, then, is that specialization in the production of livestock induced by the growth of output markets is also entirely consistent with enclosure of open fields, in the sense of consolidation of scattered

---

[12] It is a common proposition in the literature on enclosure that the size of the average farm increased after enclosure : the usual argument is that, since it is costlier per acre to fence a small farm than a large one, and enclosure required fencing, one should expect to see larger farms economically more viable after enclosure than smaller ones. While not being inconsistent with this line of reasoning, the argument in the text relies on the nature of transaction costs, in keeping with the general approach of this essay.

holdings and the abolishment of collective ownership and control.

In this way we seem to have found an explanation for the occurrence of enclosure as a wealth maximizing alternative to the open field system that can account both for the long time enclosure took to accomplish and for its wide geographical dispersion. It remains to be seen whether this theory of enclosure also accords with the empirical facts. That discussion is left for the next section of this chapter. What remains here is to emphasize the intimate relationship between conditions of production technology, the structure of transaction costs, and the influence of market conditions in determining the choice of institutional organization engineered to achieve the best possible use of scarce economic resources. In the preceding chapter the interaction between technology of production and the structure of transaction costs was stressed. What has been added in these last few pages is the additional insight that, for given technology and transaction costs, simple growth in the extent of the market is sufficient to alter the choice of institutional arrangements. In order to construct a theory for the choice of institutions it is not sufficient to adduce technological conditions alone, for the nature and purpose of institutions is to minimize resource losses due to transaction costs. The transaction costs must therefore be explicitly specified in the model. Yet it is also required that the market conditions for the output be specified in addition; these may be of such a nature as to make the assumptions about technology of production and about transaction costs irrelevant for the purpose of deciding on the best institutions. It is precisely this which has been shown in the preceding. We shall return to this observation in the concluding chapter.

The course of enclosure

If students of the open field system have encountered difficulties in understanding scattering and communal ownership, this is child's play in comparison with the complications looming around the concept of enclosure. It is simply not possible to find *the* cause of enclosure, nor will any one explanation ever account for the diversity of the process. The problem here is of a different nature from construing a model of the open field system itself.

For even if we could agree on the salient features of such a model – for example, the one presented here – it does not follow that we shall also agree on the causes of enclosure. If anything is shown by all the available studies of enclosing villages and regions it is that there is no such thing as a 'representative enclosure' that we can use as an average around which everything else turns (in the way we can perhaps allow ourselves to deal with the institutions of the open field system). Anyone who advances a strict mono-causal theory of the disappearance of scattered strips and open fields can do so only at the expense of neglecting an important body of empirical research. Consider the following statement by Yelling :

> To take only a few examples, enclosure has been attributed in par-
> ticular cases to population growth, to population decline, to the
> presence of soils favorable or unfavorable to arable, to industrial and
> urban growth, to remoteness from markets, to the dominance of
> manorial lords, to the absence of strict manorial control, to increased
> flexibility in the common-field system, and to the absence of such
> flexibility.[13]

Not only is it impossible to find *the* cause of enclosure in this mass of explanations, but there is the further difficulty that some of them are totally contradictory. We could of course dismiss them completely, and argue that they are due only to historians' over-fertile imaginations, and that only a logically tight economic theory can help us discriminate between what is relevant and what is not in such a diversity of causes. It is quite clear that this would do serious injustice to the fruits of historical scholarship. An alternative approach is to grant that it is not at all impossible that some of these albeit seemingly contradictory reasons actually did in some cases constitute the reason and motive for enclosure. Taking this approach, it is necessary to develop a theoretical framework that is flexible enough to accommodate all or some of these variations. We must therefore accept the complexity and the apparent contrariness of the observations on enclosure, and put the burden of adjustment on the theory rather than the data.

However, to apply theory or economic models, some measure of generalization must necessarily be attained, for if the emphasis

[13] Yelling, *Common Field and Enclosure*, p. 3.

on diversity and complexity is too strong, the conclusion inevitably emerges that no theory is capable of explaining the observations. If that is the case, we should agree to abandon any attempt at conceptual explanations of the enclosure movements, and be content with studying local phenomena in order just to record the past, not to explain it. Since virtually nobody takes this position, the implication must be that there is at least an implicit agreement that there are certain features that allow for a measure of generalization to be established, even with respect to such a complex phenomenon as enclosure. Indeed, there appears to be such a consensus on a few broad characteristics of the disappearance of the open fields. Even if there is no one cause of enclosure, perhaps it is therefore possible to find a *class* of causes that have something in common. It is perhaps also possible to find that some cause or causes were stronger at some times than at others, so that it is useful to structure the problem of enclosure into certain periods and geographical regions.

In the literature on enclosure, major attention has been focused on the two significant outbursts that stirred most of the controversy at the time. The first was the so-called Tudor enclosures, and, because of the alleged effects on employment opportunities in agriculture, several attempts at legislation were made in order to avoid the disrupting consequences of depopulation of the countryside. Most of these attempts failed miserably. The second was the wave of enclosures associated with, but not exclusive to, the private Acts of Parliament in the second half of the eighteenth century, leading up to the general legislation on enclosures in the beginning of the nineteenth century. These, however, are only the two major waves of enclosure, although they have captured most of the attention of both contemporary and modern writers. It should not be forgotten that enclosure is a concept which has a history much older than the Tudor enclosures. For example, the Statute of Merton is often taken to imply that the lords had the right to enclose parts of the waste :

Also because many great men of England (which have infeoffed Knights and their Freeholders of small Tenements in their great Manors) have complained that they cannot make their Profit of the residue of their Manors, as of Wastes, Woods and Pastures, whereas the same Feoffees have sufficient Pasture, as much as belongeth to their Tenements; it is provided and granted, That whenever such Feoffees do bring an Assise of *Novel Disseisin* for their Common of Pasture, and it is knowledged before the Justicers, that they have as much Pasture as sufficeth to their Tenements, and that they have free Egress and Regress from their Tenements unto the Pasture, then let them be contented therewith.[14]

The Statute of Merton was issued in 1235, and it is taken to mean that the lords have the right to enclose parts of the waste as long as they leave enough pasture for their tenants. It is to be noted, however, that this is not really enclosure in the full sense of the word, but merely an establishment of the right of the lord to those parts of the commons which are 'not basically necessary' for the tenants. It does not imply either consolidation or extinction of communal rights. Indeed, it can be interpreted as a re-affirmation of the communal rights in the waste, for the lord is allowed to appropriate part of the commons only as long as he does not infringe the rights of the freeholders on the manor. Nothing is said about the villein tenants, for the relationship between the lord and his villeins is a matter not for statute law but for civil contractual agreements to be dealt with in manorial courts.

Even so, enclosures were known from a very early time, even if we mean by that term something more than just the existence of closes in the open fields that were tilled by individuals apart from the communally decided cropping rotations with the consent of the court.[15] Yet the pace of enclosures was slow and very gradual up until roughly the middle of the fifteenth century, which marked the starting point for the so-called Tudor enclosures. Little has been written on these early enclosures, and the evidence is so scanty that it is difficult to get an overall view of the role of enclosure before the great wave in Tudor times. For example, it is at present not clear whether some areas were enclosed so early that no written records have been preserved, or whether the written records are missing because these areas never had the open field system at all. Such is the case with respect to Kent, to

---

[14] O. Ruffhead, *Statutes at Large*, London, 1743, vol. I, p. 17.

[15] B. K. Roberts, 'Field systems of the west midlands', in Baker and Butlin, *Studies of Field Systems*, pp. 229–30, exemplifies this by the dissolution of the Arden field systems as a response to market growth.

Cornwall, and to some of the upland regions, where the topography is hilly enough to prevent the laying out of extensive open fields. Yet it has been said that there were no regions of England that did not know the open field system, even though it was clearly dominant only in a broad band across England from the northeast to the southwest from the times when written records start. The history of enclosure outside these areas has yet to be told adequately.

The story told about the Tudor enclosures which has the longest tradition in the literature is that there was a sharp increase in the demand for wool owing to the growth of the Dutch wool manufactures, and that the price of wool rose very steeply as a result.[16] This led to much enclosure for the purposes of increasing the production of wool, and consequently large areas of tilled arable land were turned into pastures.[17] As a consequence, the amount of employment available in the agricultural sector diminished sharply, and some areas of the country became depopulated.

There can be no doubt that there is much basic truth in this story, even if the details need clarification and correction. The contemporary literature is full of pamphlets expounding on the human tragedies involved in the unexpurgated greed of the enclosurer, and the villain is uniformly the farmer enclosing for pasture – it would be a mistake to overlook the important implications of this literature even though its moral indignation may be responsible for some exaggerations. The importance of depopulation is corroborated by the stream of legislation flowing from the Chancery with the intent to control and halt the evils of turning people off their farmsteads. Yet the forces working for

[16] For discussions of the wool trade, see e.g., M. M. Postan and E. Power, *Studies in the English Trade in the Fifteenth Century*, New York, 1966, *passim*; E. M. Carus-Wilson, 'The woollen industry before 1550', in Pugh, *History of Wiltshire*, vol. II, *passim*; H. van der Wee, *The Growth of the Antwerp Market and the European Economy*, The Hague, 1963, vol. II; P. J. Bowden, *The Wool Trade of Tudor and Stuart England*, London, 1962.

[17] See e.g., F. J. Fisher, 'Commercial trends and policy in sixteenth century England', *Economic History Review*, vol. X, 1940, reprinted in E. M. Carus-Wilson (ed.), *Essays in Economic History*, vol. I, London, 1966, *passim*; P. J. Bowden, *The Wool Trade*, *passim*; P. Ramsey, *Tudor Economic Problems*, London, 1965, pp. 24–5; R. H. Tawney, *The Agrarian Problem of the Sixteenth Century*, London, 1912, pp. 195–7.

enclosure were so strong that no legislation could ever hope to halt it, much less reverse it.

However, the data available on the relative price of wool to grain do not seem to bear out the postulated increase in wool prices. Many people have noted that the price of wool actually fell somewhat in relation to grain from the middle of the fifteenth to the middle of the sixteenth centuries,[18] and that this flies in the face of an explanation of the Tudor enclosures based on the increase in the wool trade. The story is more complicated than this, though, for there can be no doubt that there was an increase in the production of wool at the time. But there is no particular reason to expect a uniformly increasing relative price of wool to grain. In fact, a slightly decreasing price is quite consistent with the postulate of this essay of increasing returns to scale in the production of livestock. Apart from this, however, there are some further pieces of evidence that allow us not to attach too much weight to the fact that the price of wool went down in relation to grain while at the same time there was an increase in the production of wool.

One such piece of evidence is that sheep are a joint output: if the mutton can be disposed of in the market, the total return to the production of sheep may well have gone up even if the price of wool went down. What is required is an expansion of the market for both of these joint products. Again, there is ample evidence that the growth of non-agricultural population centrally led to an increase in the demand for mutton. In addition, the fact that at a very early date wool was cheap to transport relative to other products[19] is an important element in accounting for why it was precisely in the production of wool that English agriculture specialized so early.[20]

---

[18] Harriet Bradley, 'The enclosure in England: an economic reconstruction', *Columbia University Studies in History, Economics and Public Law*, vol. LXXX, no. 186, 1918, p. 36, has pointed this out; see also J. S. Cohen and M. L. Weitzman, 'A Marxian model of enclosures', *Journal of Development Economics*, vol. I, no. 4, February 1975, p. 318.

[19] See Gonner, *Common Land*, p. 123.

[20] Perhaps an additional reason why the earlier enclosures tended to specialize more in livestock than in grain is to be found in the lack of alternative methods of fertilizing the soil: any area suitable for arable would therefore more or less be forced to keep a substantial amount of livestock until the

Yet the story of Tudor enclosures is not the story simply of sheep and depopulation : later historical research has amassed evidence for a process that is much more complex and diverse. Engrossing, that is the consolidation of two farms or more into one single farm, posed a problem apart from enclosure for pasture;[21] for when two farms were joined, the buildings on one of them were left to decay, and depopulation would follow. Furthermore, enclosure was not for sheep alone; the output of other livestock increased as well. Joan Thirsk distinguishes two broad regions where enclosure occurred in the sixteenth century : the claylands in the Midlands where much open field land was turned into pastures for fattening of cattle and for dairy farming, and the less fertile uplands where the tenants were few and the opposition to enclosure weak, so that the small open fields and the large commons could be enclosed for sheep grazing.[22] It is quite likely that further research into the details of the enclosing villages in the fifteenth and sixteenth centuries will give evidence of a process that is much less dominated by sheep, mutton, and wool than the traditional story maintains. This is especially the case since the period was one of rapid growth of population, most markedly of non-agricultural population, so that the scope for increased specialization in agricultural production was greatly widened in this period. Markets grew in scale at this time, but were still localized in the sense that they did not encompass the whole country.

It is extremely difficult to make even an approximate estimate as to the actual amount of acreage affected by the Tudor enclosures.[23] Such estimates have been attempted for the later

---

newer methods for keeping the fertility of the soil up were developed in the seventeenth and eighteenth centuries.

[21] See Joan Thirsk, *Tudor Enclosures*, Historical Association Pamphlet, no. 41, London, 1959, p. 10.

[22] Joan Thirsk, *Tudor Enclosures*, pp. 14–19.

[23] E. F. Gay, 'Inclosures in England in the sixteenth century', *Quarterly Journal of Economics*, vol. XVII, 1903, pp. 576–97, maintains that about 3 percent of agricultural land was enclosed in Tudor times. Tawney, *The Agrarian Problem*, has questioned this view, as have many others as well. However, McCloskey, 'The Economics of Enclosure', p. 125, contends that there is no reason to disbelieve Gay's original estimate.

enclosures in the eighteenth and nineteenth centuries, but there is very little available for the earlier ones. Here a tremendous amount of work remains to be done. The data may be available, to some degree, in archives with collections of surviving manorial court rolls. However, most of these enclosures were by agreement, so no records exist of the actual process of enclosure, and only records as to land use can be employed in the estimation of the areas affected.

After these great outbursts of enclosing activity in the late fifteenth and early sixteenth centuries, events proceeded at a calmer and more measured pace. This reflects the more gradual nature of change taking place in the factors that determined the shaping of the economy. If the Tudor enclosures were due in part to the rapid increase in the demand for wool and in part to the rapid increase in population, both elements together accounting for the dramatic changes in agriculture, it is clear that no such sweeping changes occurred in the following two hundred years. It is perhaps no wonder, then, that the seventeenth century is virtually bypassed in the literature on English agriculture : during this period political events appear much more dramatic than the economic ones.

The notable exceptions to this oversimplified assertion are the studies of technological change in agriculture that have recently come to the fore.[24] These studies indicate that many technological improvements of agricultural methods were conceived and implemented in this period. New methods were up-and-down husbandry, where the soil was subjected to cropping for a year or two then laid down for pasture for a period of several years, and floating of water meadows, which meant redirecting the flow of water from a river or brook so as to cover the meadow with an inch or two of slowly flowing water during the winter months. In this way the meadow was fertilized by the deposits from the water, and, since the meadow did not freeze when covered with flowing water, a considerable increase in the amount of grazing to be had on it was attained. Other changes at this time included the introduction of new crops such as tobacco, carrots, turnips, clover, and permanent grass. The importance of fertilizers, in

---

[24] See especially E. Kerridge, *The Agricultural Revolution*, London, 1967.

many different varieties, became more widely known, and new crop rotations were introduced and experimented with.[25]

Yet enclosure did occur – gradually and less traumatically than during Tudor times. Enclosure at this time was still accomplished by agreement between the farmers involved in the village, and the rule of unanimity more or less applied, so little controversy was displayed to the public eye. One reason for this was probably that enclosure during this period was not mainly for pasture, with the concomitant decreases in employment opportunities. Rather, enclosure was made for the purpose of increasing grain production as well as for pasture, and no net effect on the size of the labor force in agriculture may be discernible. This was probably due in part to the gradual growth of cities and towns that were able to assimilate the people moving in from the countryside without more difficult adjustment processes being called for.

Again, the detailed story of enclosure during this period still remains to be told, and perhaps the available evidence is so scanty that it will never be told in full. Yet it is an important piece of the history of the decline of the open field system : in recent years it has been alleged that the all-encompassing importance of the parliamentary enclosures in the eighteenth century ought to be relegated to limbo as a fairytale.[26] Rather, this alternative view maintains, parliamentary enclosure was simply the ratification of a series of changes that had already occurred at a time before the middle of the eighteenth century when the private Acts for enclosure of open fields started to pour out of Westminster. However this may be, there can be no doubt that the gradual progress of enclosure from the middle of the sixteenth to the middle of the eighteenth century served to eliminate a substantial proportion of the existing open field villages.

During this period enclosure spread first in the vicinity of the large towns and cities that provided the markets for the outputs from the improved system of agriculture.[27] There is also ample evidence that enclosure followed the gradual improvement of

---

[25] The best discussion of these matters is by E. Kerridge, *The Agricultural Revolution*.

[26] *Ibid.*, p. 24.

[27] See e.g., H. J. Dyos and D. H. Aldcroft, *British Transport*, Leicester, 1969, pp. 19–21.

transport facilities, which would have the effect of lowering the costs of taking the agricultural products to the urban markets.[28] This story is very clear, and no particular controversy surrounds it. It is part of the 'stylized facts' of the literature on the evolution of English agriculture.

Such are the very broad outlines of the progress of enclosure until the middle of the eighteenth century. At that time roughly half of the area under plough in English agriculture remained to be enclosed, according to various estimates[29] – again, none of these can be taken as final, for this is an area where controversy is bound to persist.[30] However, whatever the actual size of the area involved, there can be no doubt that the period from the middle of the eighteenth century to the middle of the nineteenth is one of extremely rapid and dramatic change in England's agriculture. In the middle of the eighteenth century, the open field system is dominant over vast regions; by the middle of the nineteenth, it is virtually non-existent, except in small token enclaves that have acquired an aura of being historical curios rather than an important mode of production.

Naturally, it cannot be presumed to be a coincidence that this last great wave of enclosures ties in with the changes in the whole economy that bear the name the Industrial Revolution; rather,

---

[28] Some eminent historians who stress the importance of regional differences and proximity to markets in the pace of enclosure include: E. Kerridge, *The Agricultural Revolution*, Joan Thirsk, *The Agrarian History*, especially the chapters by herself, A. Everitt and P. J. Bowden; E. C. K. Gonner, *Common Land and Inclosure*; H. L. Gray, *English Field Systems*, Cambridge, Mass., 1959; C. S. Orwin and C. S. Orwin, *The Open Fields*.

[29] McCloskey, 'The economics of enclosure', p. 124: '. . . through the statistical haze one can discern its crude outlines, and these are that of the twenty-four million acres of useful land in England (excluding Wales) some six million acres were enclosed by parliamentary act and, much more speculatively, perhaps eight million acres by private agreement after 1700. That is, it is understating the matter to say that half the agricultural land of England was enclosed in the eighteenth and early nineteenth centuries.'

[30] For example, some authors contend that the parliamentary enclosures really only served to put a legal seal on already performed, voluntarily agreed upon enclosures from an earlier date. W. G. Hoskins, *The Midland Peasant*, London, 1965, p. 249; Kerridge, *The Agricultural Revolution*, p. 24; G. E. Mingay, *English Landed Society in the Eighteenth Century*, London, 1963, pp. 99, 180–1, 184, 186; V. M. Lavrovsky, *Ogorazhivanye Obshchinykh Zemel v Anglii*, Moscow, 1940, pp. 22–4; E. L. Jones (ed.), *Agriculture and Economic Growth in England 1650–1815*, London, 1967, p. 13.

the problem is simply one of ascertaining the nature of the connection. Since the Industrial Revolution, among other things, was intimately concerned with the introduction of new methods of production, the inference has often been drawn that the changes instituted in agriculture were also of a technological nature. In the preceding we have on several occasions had the opportunity to question this view. The changes in technology, in field systems, breeding methods, crop rotations, crops, etc., that are normally associated with the so-called Agricultural Revolution were practically all already introduced in the seventeenth century, and were introduced in open field villages as well as on farms held in severalty. What is stressed nowadays instead is the enormous increase in the total output from the agricultural sector in the last half of the eighteenth century, as the size of the total population increased at an unprecedented rate, and the non-agricultural population at an even faster rate. If the technological aspects of the Agricultural Revolution have been overstressed, the commercial ones have not yet attained their proper place in the story of the transformation of the British economy in the Industrial Revolution.

The increase in total output of agricultural products was attained in two basic ways. First, there was an increased specialization in production in areas that previously had not been drawn into the network of regions that supplied the urban areas with their products. Another very important element was the intensification of regional specialization on a broad scale – in this period markets became truly national, so that the full comparative advantages in the production of certain outputs eventually became exploited by the English farmers in this period.[31] The second important factor in the increase of total output was the increase in the area under the plough. At this time, much enclosure was for tillage, in view of the high grain prices that followed in the wake of the wars and the increased demand from the growing urban conglomerations. As a consequence, large parts of the com-

[31] In this period some areas in the Midlands specialized in the production of livestock, and the areas in the south-east in the production of grain. E. L. Jones, *Agriculture and Economic Growth*, p. 37. Cornwall seems to have specialized very early in the production of livestock, and Middlesex enclosed at a very early date to concentrate on the production of dairy products for the London market.

mons were ploughed up, and pasture was thus turned into arable in many areas.[32] In others, as had happened before, there was an increase in the production of livestock at the expense of arable, with the now familiar effects of fewer employment opportunities and depopulation.

Perhaps the most fascinating element in the last great wave of enclosures therefore appears to be the relationship to the rest of the economy : how the agricultural sector was transformed into a network of regional specialization on a broad scale, where the whole country was tied together in a system of production, transportation, and consumption. Is it perhaps possible that the role of agriculture in the Industrial Revolution was to supply capital for both long and short term investments in infrastructure and manufacturing, and that enclosure here played a crucial role? We do know from various studies that enclosure at this time was a sometimes immensely profitable investment, and often gave good returns within a short span of years after the enclosure award. In this way vast amounts of capital could conceivably have accumulated in the hands of enclosing farmers, and these funds could have found their way into other sectors of the economy.[33] The hypothesis, for that is all that it can be at this stage, would then be that enclosure was an investment with a comparatively short term to maturity that yielded high rates of return. A question for further research is to study the role that the emerging country banks played in funnelling these funds to alternative investments.

However irrelevant these speculations may be to the issue at hand, it is clear that enclosure of open fields is a process which can never be completely isolated from events in other sectors of the economy. No matter whether medieval and early modern farmers were 'profit maximizers' or not, it is clear that they did respond to economic incentives as given to them in terms of changes in relative prices, increased market opportunities, better

---

[32] J. D. Chambers and G. E. Mingay, *The Agricultural Revolution 1750–1880*, London, 1966, p. 35.

[33] In discussing the various sources of capital during the Industrial Revolution, Flinn lends some support to the proposition that the agricultural sector indeed formed an important supplier. See M. W. Flinn, *Origins of the Industrial Revolution*, London, 1966, ch. III, 'The financial origins'.

methods of husbandry, new crops, and improved techniques for preserving the productivity of the soil. No farmer could completely insulate himself from markets, and so he was forced to respond in some degree at least to all the major changes in the national economic system that reached him. And it is therefore by looking at these major events in the growth of the British economy that we can begin to see some periods that can reasonably be isolated as units within which we can find some important general influences working towards enclosure. In the preceding we distinguished three such major periods, and this appears to be consistent with the modern treatment of specialized observers. The first is the Tudor enclosures, mainly due to the increased wool trade; the second, the period up until the middle of the eighteenth century when enclosure took a somewhat slower course; the third, a second major wave of enclosure from the middle of the eighteenth century to the middle of the nineteenth.

If we accept this threefold periodization as the generalization that describes (however inadequately) the gross outlines of the enclosure of open fields, can we show that the theory of the open field system and its scattered strips presented in the preceding chapter conforms to the broad facts of these periods? To do this we have to show (i) that an increase in demand for one of the two outputs, notably livestock in this case, would predict enclosure as the wealth maximizing solution for the farmers in the open field system; (ii) that an increase in demand for *both* major classes of output, livestock and grain, also predicts that enclosure will take place; (iii) that some changes in technology may be radical enough to spell the demise of the open field system. We have dealt with these issues already in the preceding section. It was shown there that if there is a sufficient increase in demand for either of the two outputs, we should predict specialization in that output, and enclosure, for the alternative use of one input in two different occupations with different optimal scale of usage would then be unnecessary. We would predict a gradual increase in the production of livestock until the village is in the hands of one or a few graziers that would practically eliminate the arable, at least for the purposes of producing cash crops. It would therefore seem that the model is consistent with the broad outlines of the enclosures during Tudor times. Secondly, we have shown that when demand increases for both outputs, then we should see more and more villages with diverging soil qualities drawn into the area of influence of the same market. The consequence would be that there should be a tendency towards specialization on that class of outputs which use those inputs in which the particular village has a comparative advantage. For the purposes of the individual village, the situation would therefore look very much like an increase in the demand for only one of the two outputs, the one in which it has a comparative advantage, and for the same reasons as before we should predict that enclosure ought to result. There is much in this story that is consistent with the slower course of enclosure during the period that intervened between two major outbursts in the fifteenth/sixteenth and eighteenth/nineteenth centuries. Thirdly, we must show that at least some technological changes yield enclosure. This is a more complex and involved issue. In the preceding we have argued (i) that no one technological process could be responsible for the demise of the open fields, for if that were the case enclosure ought to have been rapid and all-encompassing instead of drawn out almost interminably; and (ii) that there were several kinds of new techniques that could easily be adopted in the open field system itself, and indeed that some of them perhaps were uniquely suited for that system so that the adoption of them may even have strengthened the system and made it less prone to enclosure.

This allows us to bypass the issue that has become a bone of contention in recent literature: whether the Agricultural Revolution occurred in the eighteenth century, as traditional doctrine would have it, or whether it occurred in the sixteenth and seventeenth centuries, as the revolutionary orthodoxy would have it. It is quite possible that many of the early techniques that were introduced in the open field system gave little incentive for enclosure. At the same time it is equally possible that several later techniques indeed were not compatible with the open field system but required enclosure to be adopted. In this manner we can maintain the consistency of the model presented here with the outlines of the enclosure movement of the eighteenth century: it is probably a combination of increased market influences with ensuing regional specialization and certain new technologies that required enclosed farms to be efficient. Since we have maintained throughout that any change in techniques that upset the balance of the open field system ought to have had sweeping effects, the rapid course of enclosure in the eighteenth and nineteenth centuries becomes quite explicable.

It is a strength of this framework that it does not require a single cause of enclosure to be the one that upset the apple cart. We have identified several separate reasons for enclosure that seem to be consistent with existing studies. In the following, we shall find the clues to yet one or two more.

Efficiency and technology

With respect to the question of efficiency, conventional wisdom has it that the enclosed farms were significantly superior to the open field villages. Usually, two pieces of evidence are invoked to support this contention. The first is the significant increase in the rents charged for the same agricultural lands after enclosure as compared with before. Obviously, if the land after enclosure was able to yield a much higher income than before enclosure, there has been a great increase in efficiency. The second is the introduction of new techniques on the enclosed farmsteads, and here the inference is immediately drawn that the old open field system was inefficient since it did not introduce these improved methods of production. Often these two elements are considered to be related, so that the introduction of new techniques is assumed to explain the increase in rents.

However, many new techniques were invented and introduced in the open field system, such as up-and-down husbandry, floating of water meadows, root crops and legumes, fertilizers, and complicated crop rotations.[34] Many writers have been able to

[34] Again, the main proponent for these views is Kerridge. He states in his book, *The Agricultural Revolution*: 'This book argues that the agricultural revolution took place in England in the sixteenth and seventeenth centuries and not in the eighteenth and nineteenth . . . of the conventional criteria of the agricultural revolution, the spread of the Norfolk four-course system belongs to the realms of mythology; the suppression of oxen by horses is hardly better; the enclosure of common fields by Act of Parliament, a broken yardstick; the improvement of implements, inconsiderable and inconclusive; the replacement of bare fallows, unrealistic; developments in stock-breeding, over-rated; and drainage alone seems a valid criterion. The failure of historians to locate the agricultural revolution has thus arisen, in

show that the open field system was in no way a technologically stagnant system of production.[35] The argument that there were new techniques introduced after enclosure cannot, therefore, rely on an assumption of improperly functioning decisionmaking and control or insufficient incentives for change : the open field system showed itself quite responsive to change both in commercial opportunities and in methods of production.[36] For these reasons alone it is impossible to contend that the technological inferiority of the open field system made for its downfall. We have also endeavored to point out that any theory of enclosure that is based on technological change alone cannot explain the slow diffusion over time and the wide geographical dispersion of the enclosure movement. It is clear, therefore, that the issue of efficiency and technological change in the enclosed farms is considerably more complicated than it is sometimes made out to be.

As markets grow, there is increased scope not only for specialization, as has been stressed above, but also for diversification in production. With increased market size, the diversity of demand patterns also increases as more and more individuals with 'unusual' demands are brought into the same commercial sphere. Hence, after enclosure, there is a benefit not only from specialization, but also from diversification in production. These elements are not necessarily contradictory, for a farmer with foresight and flexibility may specialize in different crops in different years, and may be able to change from the production of one output to the next with great ease. It is in this regard that the enclosed farms most definitely must be considered superior to the

---

part at least, from mistaken notions of what form an agricultural revolution could have taken. . . . The chief criteria to be used in assessing the agricultural revolution, then, must be the floating of water-meadows, the substitution of up-and down husbandry for permanent tillage and permanent grass or for shifting cultivation, the introduction of new fallow crops and selected grasses, marsh drainage, manuring, and stock-breeding.'

[35] See W. G. Hoskins, *Provincial England*, London, 1963, p. 153; C. S. and C. S. Orwin, *The Open Fields*, p. 161; M. A. Havinden, 'Agricultural progress in open-field Oxfordshire', in W. E. Minchinton (ed.), *Essays in Agrarian History*, Newton Abbot, Devon, 1968, pp. 156–9; J. M. Martin, 'The parliamentary enclosure movement and rural society in Warwickshire', *Agricultural History Review*, vol. XV, 1967.

[36] D. Griggs, *The Agricultural Revolution in South Lincolnshire*, Cambridge, 1966, pp. 47, 54.

open field villages : when there are changing patterns of demand, there is scope for continually shifting and adapting the production of agricultural output. Such changes would be cumbersome to accomplish in the open field village, as communal consent would be required for deviations from crop rotations, or for changing pasture to tillage and vice versa. The enclosed farms, that is to say, are not only well suited to a regime of specialization, but also to one in which it would be desirable to alter the specialized output with short notice and for a short time only. In its greater flexibility, the enclosed farm would prove superior to the open field villages – if flexibility in production really carries a premium, which would be the case as market size grew.[37]

Therefore, on the enclosed farms, one would expect to see greater experimentation with crops and techniques precisely because the enclosed farms were geared to such activities, insofar as market conditions favored flexibility in production. However, there is also another element that should be noted : the interaction between the type of ownership and decisionmaking institutions in existence, on the one hand, and the changes in technology that occur, on the other. The point is that a certain institutional organization carries a presumption, as it were, in favor of certain types of changes of technique to occur and to be introduced.

For example, it is notable how some of the new techniques associated with the Agricultural Revolution are particularly suited for the open field system. Floating of water meadows was an investment on a scale totally beyond any individual farmer : it would work only if a relatively large area could be put under water at regular intervals. It therefore required that a whole

---

[37] A. H. John says : 'High prices led to a marked extension of cultivation and to a contraction of the acreage under rough pasture and sheep-walks. The spread of better farming methods – even in a rudimentary form – and the greater sensitivity of farmers to market conditions because of enclosure, improved transport, and better means of communication brought a new degree of flexibility into husbandry in many areas. There was a marked diversification of arable crops – the increased acreage of potatoes and swedes, for example; while the use of leys of varying lengths within rotations made possible some adjustment of emphasis between animals and crops in the farming programme.' See his article, 'Farming in wartime 1793–1815', in E. L. Jones, *Agriculture and Economic Growth in England 1650–1815*, London, 1967, p. 37.

meadow, perhaps of several hundred acres, could be adapted for this new method. This required that the meadow was flattened out, so that the water could flow freely, and that the soil was arranged so that the water flowed in a zigzag pattern across the field, from one end to the other where it could be led back to the river. It also required the building of canals that would divert the flow of the river, or part of it, and the construction of dams that could be opened and closed at proper intervals. This was a venture for which the open field village was uniquely qualified, for it had the decisionmaking institutions already in existence, and could easily deal with the problem of controlling the communal use of the resources going into the undertaking and the problem of individual payment for the communal resource. It would be much more difficult for farmers in severalty to reach an agreement on the mutual arrangement of water meadows, and the transaction costs would be virtually prohibitive.

The case is very much the same for up-and-down husbandry. This is a method of convertible husbandry, but different from the one in the open fields. The open fields would be cropped for a year or two, or more, depending on the rotation chosen, and would then be put to fallow; and grazing would be allowed on it, as it also was before ploughing and after harvest. The newer methods do not pertain to the open fields, but to the waste, or the commons. Part of the pasture area would be ploughed, distributed between the peasants, and cropped for a year or two. However, for reasons of poverty of the soil itself or shortage of fertilizers, these fields could not be cropped profitably for more than a season or two, and would therefore be laid down to pasture again. They would then stay down for several years, sometimes up to eight or ten years, and be taken up afterwards and put under plough again when sufficient fertility had been restored. This is also a method which is peculiarly well suited to the open field system with its organizations designed to control communal ventures. A decision would have to be taken and implemented about which areas to put under the plough, how they ought to be distributed, how the costs of fencing, drainage, and access roads should be paid for, when to let the beasts in – all decisions that the court or the village meeting made with regularity. Again for reasons of costs in decisionmaking and implementation, such ventures would be more difficult to undertake on enclosed farms.[38]

Likewise, there is a certain presumption for certain techniques to be introduced in the enclosed farms, such techniques as would be suitable for a farm in the hands of one person only, where the side effects on others would be negligible. We have already pointed to one such element : the greater flexibility afforded by one farm, one decisionmaker, so that changes in resource allocation could come about with great ease. We find here a clue to a correct interpretation of the pervasive opinion that enclosed farms displayed so much greater efficiency and experimentation than the open field villages. If there is a presumption for certain techniques of production to be introduced in the open field system, because that system is particularly well suited for a certain type of decisionmaking and control, it would slant certain other techniques which were more suited to farms owned and controlled by single individuals. This hints at two different aspects that would account for the increased activity with respect to experimentation and change in the enclosed farms compared with the open field villages. The first is that methods of production, or crop rotations and outputs, that were well-known in the open field villages, but not particularly well suited for that environment, were left on one side for as long as that system persisted. Once it had gone, the enclosed farms could easily implement old ideas that suddenly became more viable in the changed circumstances. The second is that, after enclosure, there would be a comparatively high probability of success in experimentation with respect to methods and techniques suited to farms in severalty, since little such experimentation had occurred under the previous system. Hence, the likelihood of finding viable new techniques would be relatively high, that is, relative to the situation ruling in the open field system, where experimentation in techniques suited to that

---

[38] Yelling makes an observation that fits into this view well : 'The common fields were much more capable of adapting to change within the context of a mainly arable system than they were of adapting to more pastoral types of farming. This meant that the pressure for enclosure in areas of continuing arable land use was always easier to resist, whether it was associated with new systems or not.' Yelling, *Common Field and Enclosure*, p. 33. It may be noted that it also means that when changes occurred in the breeding of livestock, for example, such new methods might make the open fields less viable, and consequently these methods would grow better in enclosed farms.

system had a long tradition. Both these aspects can then account for the fact that there was a greatly increased experimentation on the enclosed farms, whereas the open field villages seemed stagnant in comparison. Having already exhausted all of the 'easy' changes available, the open field villages would proceed at a more leisurely pace.

It is clear, therefore, that the question of the relative efficiency of the two systems is a perfectly idle one. They were equally well suited to different tasks, and comparisons of efficiency are consequently not meaningful. The wheel may have been a great invention, but it still does not run very well on snow. For that, a sled with runners is better. The open field system was adapted to cope with the problems of producing two different classes of outputs with the same resources, under conditions of few exogenous changes and consequently great stability; the enclosed farms were adopted once specialization became profitable and greater flexibility in production was desirable. Neither would serve the purpose of the other.

The issues of timing and equity

We have attempted to show that it is not possible to account coherently for enclosures unless we can first understand what made the open field system a viable organization for such a long time. Our theory therefore concentrates on the explanation of the open field system with its peculiar mixture of private and collective ownership and decisionmaking. We have shown how this theory is consistent with the available general evidence on the progress of enclosure. In this section we shall remark on some additional matters : first, the question of the precise timing of the enclosure decision in a village; secondly, the question of equity – whether the enclosure of open fields was a means by which the rich and the powerful expropriated property belonging to the poorer classes in the open field villages.

These issues are broad, and at present not settled in the literature on the enclosure of open fields. They all have a long tradition in that the earliest scholars who assembled the facts of the enclosure movements spent much effort in trying to settle them, yet no consensus can be seen to emerge. The analytical approach based on property rights and transaction costs will allow some fresh perspectives on some of the elements pertaining to these issues, but no pretense is made that any ultimate solutions are arrived at.

We have maintained throughout that, for as long as the open field system was viable, the costs were outweighed by the benefits of that system. The costs are well-known, and consist of the costs inherent in scattering, communal ownership of scarce resources, collective decisionmaking, and unexploited advantages in specialization that may have been present. What our theory of the open field system does is to identify the offsetting benefits as those yielded by alternating husbandry, with its intensive use both of the arable and of large scale grazing areas, with the concomitant benefits in livestock production. The decision to abolish this system will then depend on two related questions: the net increase in annual production from changing to a new system, and the cost of actually implementing that alternative mode of production. The first involves a present value calculation of an anticipated income stream, the second the computation of a lumpsum 'fee for change', as it were.

Our theory of enclosure relates to the first of these elements: when markets grow, the anticipated income stream on enclosed farms increases relative to open field farms, because the hitherto unexploited potentiality of specialization can now be realized. Also, when new techniques make enclosed farms more profitable, the benefits from enclosure go up. Naturally, there is an implicit element of timing already in these calculations, for the present value of the income stream to the enclosed farm would become greater than on the open field farm at a specific point in time; that is, at such a point in time when the new techniques become profitable or when the market becomes large enough to warrant increased specialization in cash crops.[39] However, before a

---

[39] It is easy within the context of the present model to visualize a gradually growing concentration in production on one of the two major outputs in the village, until one day the production of one is so unimportant in relation to the other that the village more or less naturally encloses — perhaps many of the so-called enclosures by agreement were exactly of this nature. On the other hand, if market conditions changed suddenly, as they did during Tudor times for example, the resultant enclosing activities could be envisioned to come about quite rapidly, and force perhaps traumatic adjustments on the open field villages.

definite date can be set for the actual enclosure, the investor contemplating it would also have to take into account the second
element in the decision process : the lump-sum cost of changing.
This once-for-all cost has several components. One is the cost
of fencing, another the cost of reimbursing the parliamentary
commission that carried out the enclosure, if there were such a
commission.[40] These elements are straightforward, and could
probably be computed relatively easily. There is yet another,
probably substantial, type of cost : the compensations to various
individuals for their, perceived or real, losses from the change to
farming in severalty. These costs would be much less predictable.
For example, it is often stressed in the literature that the bribes
necessary to secure a favorable decision in parliament could be
very high. The point is that the cost of these bribes would also be
unpredictable. The same would be the case for the sum that
would have to be paid out as compensation to the poor who lost
their rights of common. This would depend on the evaluation
made by the commission, and could not be easily predicted. Also,
there must have been the problem of convincing recalcitrant
farmers in the village that they should promote the enclosure,
and this again would involve either straight bribes or promises
of increased shares in the village lands after enclosure. Lastly,
since the actual allocation of rights in the village was made by
the commission, an important element of uncertainty must have
been introduced in the estimation of exactly how much each
farmer would receive after enclosure.

Because of all these uncertainties inherent in the costs of changing to a new system, it is easy to rationalize why a neutral body
such as the parliamentary commission was resorted to in order to
decrease the costs of enclosing. Without the commission, the costs
would have been even higher, since the distribution of assets in
the village would have had to be solved by straight individual
bargaining, and much time and resources would have been
wasted in this process. The commission could not eliminate all
such costs, of course, but must have been able to reduce them

---

[40] M. E. Turner makes the observation that these costs often were much more
substantial than has previously been believed. See his 'The cost of parliamentary enclosure in Buckinghamshire', *Agricultural History Review*,
vol. XXI, 1973, pp. 35–46.

substantially. Even so, because of the transaction costs involved, it is readily seen that it is impossible to determine a simple criterion that will yield a single date for carrying out the enclosure: we have attempted to show that the costs of changing to the new system will be determined, partially, by the specific individual characteristics of the transactors involved in the decisionmaking procedure. If any one farmer holds out against enclosure, he can make it very costly indeed for the rest to achieve the benefits from enclosure and specialization; the costs could, on the other hand, have been very small if it happened that everybody from the outset agreed on enclosure.

The net benefits from enclosure, and the precise timing of it, are therefore no longer feasible to calculate with conventional cost–benefit criteria, for implicit in the nature of transaction costs is an element of indeterminacy in the extact value of the costs and the benefits. In Chapter 3 we discussed various notions of transaction costs, and it was shown how transaction costs cannot be treated as general phenomena, but must be regarded as individual-specific. This means that any theoretical analysis of enclosure must recognize the power of each individual to make a nuisance of himself, and thereby impose costs on others from which that individual himself can derive benefits. Until we have specified the individual make-up of the persons involved in the enclosure decision, we cannot calculate the net gains from enclosure. The conclusion is that, for very straightforward theoretical reasons, it cannot be expected that enclosure would, for example, spread concentrically around an increasing market area. For an ambiguity in the element of timing is here introduced that does not rely on the standard arguments of inertia or conservatism among some peasants and villages in relation to others. In this manner we can explain why enclosure is a process with a strong random element, and this may account for part of the variations in timing, as well as the wide geographical dispersion of enclosure.

The next issue to be discussed here is one that has captured a great amount of attention among the scholars who have made the enclosure of open fields their special domain. It concerns the problem of equity in the enclosure of the open field village. Who lost and who gained? How was income distribution affected? Did the poor come to suffer even more from enclosure than they already did in the open field village? This is one of the most hotly debated issues pertaining to the open field system and the enclosure movements. No solution will be offered here; all that will be attempted are some remarks that follow naturally from the transaction costs approach taken in this study.

One of the pieces of evidence often adduced to show that the small and less wealthy farmer stood to gain little from enclosure is that the main promoters of enclosure were supposedly larger and more wealthy. The main reason for this is thought to be the costs of fencing: it is natural that it costs more per acre to fence a small holding than a larger one, and since fencing was a necessary corollary to enclosure, the small farmer stood to gain relatively less than the larger one would. Sometimes the argument is put that small farmers actively resisted change to farming in severalty, but that they were overriden by their more powerful and wealthier neighbors. There are some empirical observations that would support such an interpretation, it seems. One is that the size of the average farm increased after enclosure. That is, the small farmers sold out after enclosure.[41] Obviously, or so it is inferred, they could not afford the cost of enclosure, and came to suffer from it, since they were driven away from their holdings. Another is that parliament has recorded many petitions from smaller farmers who resisted enclosure, and certainly they would not have protested if they had stood to gain from the change.

The interpretation of the available evidence is by no means clear-cut, and much historical research remains to be done before it will be possible to show unambiguously that the small farmer actually disappeared with the enclosure of open fields. In some cases it appears that the size of farms did not increase significantly after enclosure, so the argument that the small farmer was evicted has less force than has often been attached to it.[42] Also, it is not

[41] See e.g., Hoskins, _The Midland Peasant_, pp. 165, 249–51.

[42] This is the contention of M. E. Turner in his 'Parliamentary enclosure and landownership change in Buckinghamshire', _Economic History Review_, 2nd series, vol. XXVIII, no. 4, 1975, p. 565 _et seq._ Turner notes that it seems that not only small farms, but indeed farms of all sizes were changing hands after enclosure of a village. He shows with evidence from the land tax assessments that there was no particular tendency for the number of small farms to decline. However, it is to be noted that the land tax assessment only covers owners. Until we know what happened to the number of

at all clear that it was exclusively the larger farmers who promoted enclosure, as there are many instances of smaller farmers also being actively involved in working for enclosure. These observations have to be balanced against the ones that seem to speak for an opposite interpretation. However, let us grant, for the sake of the discussion, that the size of the average farm actually did increase after enclosure, and that many small farmers sold out and either became farm laborers or moved to urban areas to work in other occupations. Even under these circumstances, must we conclude that the smaller farmer actually did come to suffer from enclosure, and that some of his wealth was appropriated by the larger farmers imposing enclosure on him? And if that was the case, how does this relate to the previous argument of the purported efficiency of the open field system?

The first point to note is that, before the machinery for achieving enclosure through a private Act of parliament came into use, unanimity, or at least virtual unanimity, was required before enclosure could be accomplished. An individual farmer's rights of common could not be abolished without his explicit consent. Hence, even if there was only one recalcitrant farmer in the village, enclosure could not come about. The only exception to this is the Statute of Merton, which, as pointed out above, allowed the lords to enclose part of the waste if sufficient pasture was left for the peasants.

With respect to the freeholders, the situation was this:

. . . the freeholder's right of common is his several right, as much his several right as is his tenancy of his house. His 'seisin' of this right is fully protected by the king's court, protected by a similar action to that which guards his seisin of his house; . . . The individual freeholder addresses his lord and his fellows : – 'True it is that the waste is superabundant; true that I am only entitled to turn out four oxen on it; true that if half of it were enclosed I should be none the worse off; true that all of you wish the enclosure made; true that I am selfish : – nevertheless I defy you to enclose one square yard; I defy you severally; I defy you jointly; you may meet in your court; you may pass what resolutions you please; I shall contemn them; for I

---

farms leased to tenants, and how the size of these compared with pre-enclosure farms, we cannot really tell what happened to the average size of farms.

have a right to put my beasts on this land and on every part of it; the law gives me this right and the king protects it.' This is not communalism : it is individualism *in excelsis*.[43]

There are numerous instances on record where the freeholders upset the attempts of the lord of the manor to enclose the waste or the open fields. The ability to do this did not rest exclusively with the freeholders, for the copyholders had well defined rights as well, and the lord could not simply ride rough-shod over them as he pleased. The rights of the copyholders were guaranteed by the custom of the manor, and if the custom of the manor said that there were rights of common appendant to certain holdings, the lord could not disregard this. The rights in the land belonging to the copyholder were his by right of a contractual agreement between himself and the lord, and as long as that contract was in force, the lord could be guilty of disseisin if he enclosed against his own tenants, and an action could be taken against him. If the lord wanted to enclose his lands he would have to wait until the leases of his land ran out, and then he could accomplish what is called enclosure by unity of possession, i.e., since the lord would no longer have tenants he could do as he pleased with the land.

The point here is that there appears to have been adequate safeguards against the larger farmers exploiting their greater economic power and simply expropriating some of the wealth belonging to their smaller neighbors. As long as the freeholders opposed enclosure, it could not be accomplished; and if anyone then opposed it, his cooperation would have to be secured by making sure that he did not lose from the enclosure. In other words, his cooperation would have to be purchased through suitable compensation.[44] As long as this was the situation it seems reasonable to conclude that the owners of the land at least did not stand to lose from enclosure.

---

[43] F. Pollock and F. W. Maitland, *The History of English Law*, Cambridge, 1968, vol. I, pp. 262–3.

[44] 'Thus agreements real or fictitious were secured. Probably where but few were concerned it was not difficult to bring people to a voluntary assent, and in other cases by mingled cajolery and pressure dissent could be prevented. But the complexity of rights which existed in the larger number of open fields and the growing knowledge that decrees obtained in chancery did not bind a dissentient minority rendered resort to parliamentary sanction desirable.' Gonner. *Common Land*, p. 182.

The case is a little different with respect to enclosure through Act of parliament. In these cases it was usually enough if a majority of three-quarters or four-fifths, reckoned either in area or value, voted for enclosure.[45] In such a case the larger owners would have a greater influence over the outcome than the smaller ones, and it would seem that they would be in a position where they could take advantage of this. However, their ability to do so was limited by the commission set up by parliament to effect the enclosure itself. This commission would determine the proper compensation to those who stood to lose from enclosure, and the equitable distribution of the costs of the enclosure.[46] It is often remarked that these commissions carried out their difficult task with reasonable success.[47]

Quite apart from these considerations, it is easy to show that even if the average size of farms increased after enclosure this cannot by itself be taken to imply that the small landowner necessarily lost from enclosure; the matter is considerably more complicated. Two points should be made. The first is that even if the rules of the game were set up perfectly equitably, so that no farmer lost anything by the enclosure itself, it is still possible that some smaller or larger farmers may have found it profitable to

---

[45] 'The parliamentary committee of 1800 on enclosures reported that there was no fixed rule; that in some cases three-fourths was required, in others the consent of four-fifths, but this refers not to the number of persons but to the value of the property, calculated sometimes in acres, sometimes in annual value, sometimes in assessment to the land tax, sometimes in assessment to the poor rate. Thus one owner who possessed three-fourths or four-fifths, as the case might be, could override the wishes of all the others; but there is little trace of this in the Acts which usually contain a goodly number of names.' W. H. R. Curtler, *The Enclosure and Redistribution of Our Land*, Oxford, 1920, p. 153.

[46] McCloskey, 'The economics of enclosures', states: 'The usual majority required in the early years of the procedures was four-fifths of the land of a village, voted by its owners, a formula that is itself significant: under both the common law and equity an owner of any rights in the open field, a category which included tenants owning long leases (sometimes even those owning yearly leases) and cottagers owning minor rights such as that of gathering the gleanings of the harvest, was considered an interested party whose veto was final.'

[47] 'When the gravity and the delicacy of the task undertaken by the commissioners was considered, the existence of complaint against them is not astonishing. It is rather a matter for wonder that the complaints were not far louder and universal.' Gonner, *Common Land*, p. 82.

G

sell out and move away. Suppose that for a typical farmer there is a net increase in rents, and hence in the value of the property, after all the costs of enclosure have been accounted for. Such a farmer would naturally stand to gain from the enclosure. Yet it is quite conceivable that the value of that same farm increased even more if it became attached to another farm after enclosure. In this case the owner would sell out, even if he gained from the enclosure itself, for by doing so he could get two capital gains : the increased value of the farm from being enclosed, and the differential rent compared to the other farm to which it would become attached when sold would become additionally capitalized into the value of the first farm. Such would be the case, for example, if there were appreciable returns to scale that could only be realized through enclosure. The cost of fencing can also be considered of some importance in this context. It has already been pointed out how the cost of fencing decreases per acre with the size of the farm. Hence, there is some foundation for the belief that in some cases there was an economic reason for expecting an increase in the size of farms. However, it does not follow that whoever sold out also lost.

This is not to say that there were no losers from enclosures. It is quite possible that there were substantial losses incurred by small owners, but the fact that small owners sold out is not a sufficient argument to establish the fact that all small owners actually lost from enclosure. Note also that an owner who did lose might not necessarily sell out after enclosure, for it is not certain that his land was worth more after enclosure to someone else than it was to him. That is to say, enclosure may conceivably impose a once-for-all capital loss on the small owner, without there at the same time being an increase in the value of the land in the hands of somebody else. In this case the land would stay with the previous owner, and no transaction would take place. As long as the net capital loss imposed on the owner is less than the increase in value of the land if sold to somebody else, there would be a net loss for the small owner, and no exchange of the land would take place. Clearly it is quite wrong to suggest that, simply because smaller owners sold, they also lost from enclosure. Some who sold may have lost nothing while some who never sold may have lost considerably.

Nor is it possible to maintain that because many small owners lodged complaints about the prospect of enclosure, or publicly opposed enclosure, that they actually came to lose from changing to the new system. This problem pertains to the nature of the decisionmaking process under a rule of unanimity. On the one hand, this rule is expressly constructed so as to ensure that no one actually comes to lose from the collective decisionmaking process, while on the other, it considerably increases the costs of actually attaining the changeover to a new system. For under the rule of unanimity excessive power is vested in any one participant, who, although he may actually stand to gain from the change, can use his power to disagree with the majority to extort the economic benefits accruing to the others from the change. By refusing to go along with the proposed change, one single owner puts himself in a position of monopoly *vis-à-vis* the rest, and can, in principle, force them to pay all their gain from the proposed measure to him.[48] It is not clear that many of the complaints registered by those who opposed enclosure were not of this nature: an attempt to capture a greater share of the increase in income than they would otherwise be entitled to.

It is in this light that the latter legislation of eliminating the necessity of unanimity must be seen. Powerful economic forces were working for enclosure, and it could not be tolerated that the unscrupulous few would be in a position to impose such high transaction costs on the rest of the community that the benefits from enclosure actually could be substantially diminished. It is quite understandable that the rule of unanimity was abolished, and that the commission set up by parliament was given the task to act as a neutral body in the cases of litigation among the tenants in the open field villages. The complaints against the parliamentary commission invites an interpretation similar to the one with relation to the rule of unanimity: by lodging strong complaints, and by being insistent, it is quite likely that individual tenants could end up obtaining a better deal, even if they stood to lose nothing from the enclosure itself. Again, it is not clear that the existence of many complaints in any way warrants the conclusion that enclosure constituted robbery of the poor.

[48] McCloskey, 'The economics of enclosure', pp. 132–3, makes a similar argument.

G*

This interpretation explains the fact that such complaints against enclosure are concentrated in the two major periods of enclosure : the Tudor ones, and the enclosures in the eighteenth century. For the more slow and gradual the change to the new system is, the more easily can its effect be foreseen and contended with. For example, certain persons in the open field village must have lost from enclosure, such as the persons hired directly by the local community to perform tasks for the court or the village council. They no longer had any role to play after enclosure, for their professions were no longer necessary. And such was the case for the beadle, the reeve, the herdsman and the shepherd. If enclosure came about slowly it would be possible for these people to adjust gradually to the necessity of finding a new profession, and they would not suddenly find themselves destitute. The same is the case if there was a stubborn farmer who insisted that he would use his potential monopoly power under the unanimity rule and appropriate more than his share of the gain. He could be waited out, and when he died or his lease came up for renewal, he could be bypassed. The more urgent the change is, the more scope there will be for certain individuals to impose costs on the others, and to gain for themselves thereby. It is quite natural, then, that the intervening period is relatively free from great complaints, whereas the concentrated outbursts of enclosing activity in the eighteenth century and in Tudor times have received so much attention. As long as enclosure went ahead at rather leisurely a pace, the victims would have ample time to see it coming, and could find new occupations and move out in good order. The greater the intensity of the exogenous disturbances that precipitated the changes in organization of agriculture at the time, the more traumatic the consequences for the individuals concerned, and the greater the possibilities for some to impose transaction costs on others. Much more research remains to be done before it is possible to ascertain exactly who gained and who lost from enclosure.

The claims for the theory

In discussing alternative explanations for the open field system and the enclosure movements in Chapter 2 above, a framework was set up for judging alternative models. This framework has also constituted the empirical material of this study : it is the 'representative village' and its salient features that we have attempted to explain in as precise a manner as possible. It will be recalled that the major criticism levelled against previously advanced theories of the open field system was that they failed to account simultaneously for all the stylized facts in the description of the open field system and the enclosure movements. The time has now come to evaluate the alternative framework presented in this study to see if it performs any better in this regard. Only if this question can be answered in the affirmative can there be any basis for the claim that the property rights approach of this study has contributed to our understanding of the open field system and the enclosure movements. We shall therefore refer back to the stylized facts of the representative open field village, and this will serve both to evaluate and to sum up the analysis presented so far.

With regards to the physical structure of a village, there are essentially three different features that require explanation. First, one must account for how the total land area is divided into arable and grazing land, and for the fact that the latter, or the waste as it is commonly known, was not subdivided. Several considerations must enter into this determination. In the long run, perhaps the most important of these is the intimate relationship between livestock and grain. In particular, since few alternative forms of fertilizers were known for most of the period of the open field system, the cattle and sheep were essential for keeping the productivity of the soil at a reasonable level. This may eventually turn out to be one of the crucial aspects in explaining the geographical distribution of the open field system : in various regions alternative forms of fertilizers were available at reasonably cheap price. This was the case in certain areas close to cities, where refuse and other matter was carted out and ploughed into the fields, or in areas close to the sea, where seaweed and various minerals were more abundant and could be used. In the short run it was of course possible for a village to respond to variations in market prices and alter the proportion of the soil that was used in either arable or grazing. In addition, population growth exerted a pressure on the arable over longer periods of time, and assarting continuously diminished the extent of the waste. In this manner it is clear that the basic determinants of the division of the land into arable and grazing areas were knowledge of technology, especially of fertilizers and alternative crop rotations that could preserve the quality of the soil, and market prices, depending on variations in demand over time.

Second, we must explain the formation of large open fields, especially in view of the fact that there is ample evidence for large-scale remodelling of many open field villages, and the introduction of large fields. The explanation we have advanced in this study is the same as for the lack of subdivision of the waste: essentially, it is the desire for large-scale grazing. The point is that if the land is kept undivided and the fields large, then the output from the livestock is greater for a given area of land. Clearly, if this is a technologically given condition, determined by the behavior of livestock and the time needed for grass to recuperate after being trampled down, then the formation of large grazing areas makes much economic sense.

Third, there is the question of the scattered strips. The explanation we have advanced here has two components: one is technical, the other relates to the incentives for selfish bargaining behavior afforded by communal grazing. That the strip and its shape ultimately depend on the technicalities of ploughing and on the topography of a particular village there can be little doubt. Furthermore, we have accepted the various reasons for the introduction of the strips that have been advanced in the historical literature, and the explanation for the degree of scattering that the risk aversion hypothesis yields. The reasons for the persistence of scattering is then dependent partly on the desire for alternating or rotating husbandry in the arable. This creates the complication of having to cope with the alternating use of the same piece of soil in two activities where the optimal scale of use is very different. The second element is that this introduces the possibility of strategic bargaining behavior on the part of some unscrupulous individuals, for by threatening to withdraw from the cultivating cycle they could thereby inflict costs on the other members of the village. In addition, the existence of the strips affords improved incentives for the farmers to participate in and keep viable the collective decisionmaking organization that regulates communal affairs. The point about the strips is that they alter incentives by changing the constraints under which each peasant acts.

In accounting for the ownership structure, we have made the assumption that the village consisted of a certain amount of land that constituted the economic assets over which the village community exercised its control. This may have given the impression that the village size was an exogenous variable, beyond the control of the community. This is true, insofar as the disputes with neighboring villages over areas of intercommoning, for example, were decided by legal matters beyond the control of the community. But to some extent, at least, the size of the village was a decision variable for the community, for it would depend on the particular output mix that the community decided on as being the most economically advantageous, as well as on topographical considerations that also determined the production decisions of the farmers. It is of crucial importance to the present analysis that we may reasonably consider the village size as given in this manner, for otherwise the free entry problem raises the possibility of the dissipation of rent. With regard to historical fact, there seems little problem in accepting the proposition that the village indeed did constitute an entity that controlled and shaped its own destiny – at least within the limits set by climate, soil quality, topography, market considerations, and the laws of king and country.

We have argued that the reason that the arable was privately owned, that is in principle controlled by private decisionmaking, was the better incentive towards efficient use of a scarce resource that is so well recognized as a central proposition of the property rights literature. With private ownership a farmer would be responsible for incorrect decisions, as well as being in a position to gain personally from good ones, so that there is a strong incentive to avoid inefficient uses of the soil in favor of ones that contribute to increases in the income for the farmer. Naturally, this is not to argue that all farmers were always inevitably 'profit maximizers', in the strict textbook sense of that word. The point is only that the institutional constraint of private control of cropping in the arable gives better incentives. It would then be up to the individual farmer to decide whether he would want to respond to those incentives or not.

This leads to the question of the communal ownership of the grazing areas, or the commons. The argument for why this is an efficient organization in the context of the production problem of the open field village is that, when large-scale grazing is desirable, private ownership gives incentives to opportunistic and selfish bargaining behavior by unscrupulous individuals. Collective ownership bypasses this problem, while at the same time affording the farmers the possibility of selling or renting grazing rights, should they not find them useful for their own particular purposes.

Such transaction cost considerations are also the key to understanding the reversion of the arable from private property to collective and back again in the well controlled cycle afforded by the practice of common of shack. Given that the incentives are superior when the arable is owned privately, there is again the complication that a private owner could withdraw his plot from communal grazing. This is avoided by imposing a servitude on the individual owner to the effect that there is a communally owned right to grazing on everyone's property during certain time periods – in effect making it collective property when used for grazing. Thus the best of the incentive mechanisms of collective and private property are preserved, so that efficiency in the allocation of resources can be attained at the cheapest cost.

With respect to the institutional structure, the preceding arguments about the relative efficiency of collective vs. private ownership in each specific instance implicitly also yield an explanation for why certain decisionmaking powers were vested either with the individual or with the community. Indeed, this follows almost automatically from the identification we have made between property rights and decisionmaking rights. It only remains to point out that there is in principle nothing that says that there should be communal decisionmaking with respect to all the bylaws passed down by the court or the village meeting, but this is really a question of political influence.

However, there is the question of the voting rule that at least implicitly seems to have been used in the heyday of the open field system and that later came to be formalized by parliament in the general legislation on enclosure passed in the first half of the nineteenth century. The democratic principle of one man one vote was never espoused by the open field village community; instead, influence stood in proportion to income or wealth. We have argued that there is an economic justification for this in that such a decision rule partly makes the influence that others exert on one man's property more palatable to that man, so that this may induce him more easily to partake in the communal organization, and partly also represents a way to circumvent the poor incentives that are associated with collective ownership and control in some instances. When the influence of each man stands in proportion to his input, then the tendency towards the emergence of a free rider problem is diminished. There is no obvious reason why we should expect the decision rule to be a democratic one. The issue at stake is not political influence and broad social issues; rather, it is one of the proper utilization of scarce resources. It is not at all clear that the village community would have functioned better under democratic rule, as seems to be implied by the arguments that, in particular, some Marxist writers advance – quite to the contrary. If the framework presented here based on transaction costs and the choice of property rights is accepted, there is every reason to believe that such a decision rule would prove very destructive, as the larger landowners would quite probably insist on going their own way to avoid the capturing of a large part of their income by the community in exercising democratic control over scarce economic assets. If that were the case, we would expect to see the villages break down altogether. The implication is that the voting rule, far from being simply an element of exploitation of the poor, indeed formed part of the very web that kept the village community viable.

The framework also accounts for two other historical facts: that there were usually some controls exercised by the community through the courts both on exchanges of strips in the arable, and on cropping rotations. We have shown how the scattered strips probably did impose costs on the individual farmer, and how he therefore would have the incentive to try to achieve consolidation, if he at the same time could preserve his communal rights in the land belonging to others in the village. In order to prevent everyone from consolidating in this manner – which would increase the cost for everyone of attaining a maximum income from the given productive resources – the community must retain some

control over exchanges.[49] That this was done is amply documented
in the bylaws. Whenever a farmer exchanged strips with someone
else, he would have to pay a fine to the court for recording the
transaction. Implicit in this is the power of the court to deny
certain transactions. There is no reason to believe that this was a
severe problem in the open field system. If everyone, more or less
consciously, realizes the benefits from scattering and the use of
the open fields for grazing, then there is little reason to believe
that everyone has to be constantly cajoled into participating in
the affairs of the community, in the interests of internal stability.
However, there may occur instances in which it is necessary to
convince some particular individual that the long run interests
of both himself and the other members of the village are best
served if complete consolidation is avoided, and in that case the
control of the village over exchanges may become crucial. This is
again an example of how transaction costs can be imposed by one
individual on the rest of the community, and how an institutional
rule is devised so as to avoid the negative consequences of that
individual's contrary behavior. In addition, the risk aversion
hypothesis notwithstanding, it indicates that there is no such thing
as a uniquely given 'optimal number of strips'. For it is natural
to expect that, from the viewpoint of the community, the optimal
number of strips for one stubborn and deviating misfit would be
considerably larger than for someone who is a pillar of the community.[50] Therefore it is likely that the power of the community

---

[49] McCloskey's, 'The economics of enclosure', p. 129, says: 'One obstacle (to
enclosure) . . . was that the lord of the manor had the right, except by
local custom in Kent, to permit or to prevent the exchange of lands among
those who held lands of him.' Many other authors stress this point.

[50] In McCloskey's theory of risk sharing, what is crucial is the number of
'real strips' as opposed to the number of 'accounting strips', as it were,
since strips in blocks could be worked together. In the model presented
here, on the other hand, it is the number of 'accounting strips' that matters,
for the individual farmer would require permission to exchange any one of
these strips, and it is the communal control which is significant for the
purposes here. Indeed, in McCloskey's model it becomes inexplicable why
the distinction was preserved at all; whereas, in the present approach, the
number of 'accounting strips' serves a very real purpose. That this number
could be very high is shown in the following little quote: '. . . in one
manor, we are told, that a tenant owned 19 acres in thirty-six different
strips, and a common field of 1,074 acres was divided among twenty-three
owners who had therein 1,238 separate parcels.' Curtler, *The Enclosure and
Redistribution*, p. 113. This makes 53.8 strips per owner and .87 acres per

to restrict the exchanges of strips was used most circumspectly, and that it was a weapon never necessary to wield over anyone in the village.

The case is different with respect to cropping rules. Such rules would be necessary in order to regulate the changeover from privately owned arable to collectively controlled grazing on the arable. In some instances this took the form of actual cropping arrangements where all the farmers sowed the same crops and harvested together. In others, such detailed regulation does not seem to have been imposed, but only dates for opening up the arable fields for grazing were determined, and the choice of crops left to the farmer. Again, this is a transaction cost problem, for in the absence of such rules it would be impossible to avoid disputes between farmers over whether A has the right to grazing before B has taken in his crops, and so on. There is no doubt that the net income for everyone in the community is higher if everyone agrees to limit his own decisionmaking power by recognizing the right of the court to pass such bylaws.

The case is the same for the restrictions on the use of the commons. Here it would naturally be in everyone's interest to agree to some control of the access to the benefits that the commons yielded. We have repeatedly referred to the dissipation of rent that otherwise would occur, and there is no reason to dwell on it again. The point is simply that private wealth maximization again compels each farmer to agree to restricting his own rights.

With respect to the technological structure, there is little to add to the preceding. We have already discussed the differences in the optimal scale of land use in cropping and in the raising of livestock, as well as the interrelationship between the two major outputs of the typical open field village. We have also noted how this implies that in every village we should expect to see both classes of output being produced, and how this leads to an explanation of the fact that the open field system seems to be characteristic of areas that produced both outputs as cash crops, as there was little specialization in the open field system. It might be observed that, although these are observations that are true

---

strip. Curtler calls this system 'mingle-mangle' – but the point is that the system must have served a purpose, or else why bother to preserve the records of actual strips?

for the village as an entity, there is no particular reason to expect that each and every farmer invariably produced both outputs. It is quite conceivable that some farmers concentrated on raising crops, and then rented out their rights to grazing, or sold them outright. It might be observed, however, that the prevalence of farmers that indeed did produce both classes of outputs may very well be due to risk aversion, for it can reasonably be expected that the price of livestock moved in the same direction, in the short run, as that of grain. The consequence would be that, as long as short run price changes were due to changes in supply, a risk averse farmer could attain a decreased variance of his income by producing both outputs.

There remains the complex question of the evolution of the system over time and space. We have argued that the open field system represents a method of utilizing the soil intensively, and that it therefore represents a response to the growth of population over time. This gives a clue to the timing of the introduction of the various elements of a mature open field village, as we have shown in the preceding. However, it also points to explanations for why there are so many observations of incomplete or immature forms of the system : scattering without grazing of the arable, examples of villages where farmers held strips in only one field, use of commons that were not stinted, consolidation through exchanges, no communal rules of cropping.[51] We now interpret these incomplete forms of the system as stages along a dynamic adjustment path, on which the community tries out various alternative solutions, discards those that do not work, copies from other villages those that do, so that all villages, in response to similar influences, had a tendency at least to evolve towards a

---

[51] Yelling makes an observation that also seems to contradict the tenets of this study : 'Common rights might be abolished, and yet the land lies open and intermixed; or alternatively land consolidated and enclosed by hedges might still be subject to common rights. Moreover, consolidation, abolition of common rights, and physical enclosure were not events that necessarily or even usually took place at a single moment in time. Each could be achieved in a gradual manner.' Yelling, *Common Field and Enclosure*, p. 5. However, as long as such deviations from normal events were not too common, so as to make them the norm, there is really nothing in these observations that should make us reject our theory – we still cling to the representative course of events, and will not feel obliged to account for every deviation history manages to come up with.

common solution : a mature open field system. Therefore it is really not inconsistent with the analysis presented here that there are individual examples where not all of the elements of a typical village are observed. Apart from the fact that these may represent villages on the way to imposing a mature system, it is again useful to point out the fact that we do not require the theory to conform to every known instance : it is sufficient if the theory does explain a central tendency that we see at work. We are concerned with the representative village, not with all the little variations around its basic theme.

Since the open field system represented an intensive use of the soil, such as would be necessary to impose in response to growing numbers, it is not difficult to reconcile the fact that the open field system disappeared in enclosures at the same time that it was imposed in other parts of the country. It is quite likely that, whatever the particular reason for enclosure, there came a point in some places where the demands placed on the soil became so strong that the only solution was to increase the efficiency of the use of scarce soils by imposing the open field system, as the case of certain remodellings seems to imply. The point is that the open field system indeed proved efficient in getting the most out of the soil relative to older methods, and relative to the number of people that were to be fed off the soil. Only in this manner can we ever begin to explain why the open field system was devised, imposed, and preserved at all.

In accounting for enclosure, we have pointed to several disturbances, exogenous to the village itself, that would lead to the breakdown of the internal mechanisms of an open field village. Increased demand, changes in relative prices, and technological change are the three broad classes we have identified that can account for enclosure. This makes enclosure consistent with a broad class of events that all formed an important part of the economic development of Britain from the Black Death onwards. Population growth, increased urbanization, increased export demand, changes in relative prices through income effects and wars, improvements in transportation and infrastructure, and discoveries of improved methods of fertilization and crop rotations, to name but a few, are all events that together or by themselves can contribute to an understanding of the timing as well as geographical dispersion of enclosure. In addition, this accounts for the fact that enclosure did not disband the open field system in one fell swoop.

We have also pointed to one additional element why it is reasonable to expect that there is a random component in the determinants of enclosure : the possibility of strategic behavior on the part of some members of the community. Naturally, this is a possibility that was open to many an enclosing lord, and one to which perhaps many of them resorted. In addition, however, we have endeavored to show how even the seemingly lowly could use their power to disagree to advance their own cause. In this manner an enclosure decision was quite likely delayed for long periods in individual instances. This problem is inherent in the dispute over property rights, since such rights have the two aspects of providing constraints and incentives as well as being the method for determining wealth or income distribution.

Such is the list of stylized facts that we set out to explain. Whether the theory developed is convincing or not, time alone will tell. It does not claim to be the last word on the open fields and enclosure – on the contrary, it is but a first step. Yet it is hoped that it will prove fruitful in providing directions for both new conceptual inquiries, as well as for detailed empirical investigations beyond the abilities of this writer. However, the claim for the theory is that it does tie together better than any current theory the broad outlines of the open field system and the enclosure movements in a logically consistent manner. Viewed in the light of the analysis of this study, the open field system appears to be an institutional structure that intelligently and delicately dealt with a number of intricate issues pertaining to the use of scarce resources, the implementation of various ownership arrangements, as well as some difficult income distribution problems. With respect to these tasks, it was a remarkably efficient and useful instrument; and the fact that some of the tasks became obsolete can in no way detract from the ingenuity and perseverance of those who developed and preserved it throughout the centuries. There are no dumb peasants, only inadequate theories of those researchers who attempt to explain peasant behavior.

In addition, however, to forming a consistent explanation for the stylized facts, the theory advanced here has allowed for a fresh insight into the problems of communal ownership and control. It has been shown that, depending on the structure of transaction costs, there are indeed instances in which collective ownership and control can be made quite consistent with private wealth maximization and efficient utilization of scarce resources. We have also shown how the particular problem of the open field system has forced us to develop the notion of transaction costs one step : only in the light of the realization that transaction costs are individual-specific, rather than general, can we gain an insight into the organization of the representative open field village.

6

SOME EXTENSIONS AND GENERALIZATIONS

After having presented a theory for the open field system and its demise in the enclosure movements, one task remains : to generalize that theory and see if it contains anything that can allow for some insight, however small, into the much wider problem area of the choice of institutional arrangements in general. If the economic theory of the choice of institutions in the open field system presented here has some validity, it ought to follow that we should also find clues to the correct interpretation of other similar institutions in the methodology employed. We shall see that this indeed is the case, and that the general ideas employed throughout this study have some applicability when it comes to understanding the processes by which institutions emerge, persist, and eventually become replaced with newly emerging institutions. In order to accomplish this, we shall also have to discuss the relationship between the standard theorems of the property rights literature referred to in Chapter 3. More specifically, we shall have to show how the contention that collective ownership indeed can be consistent with private wealth maximization can be justified in view of the works of Cheung and Demsetz which purportedly show the superiority of private property rights over collective.

The efficiency of collective ownership

In Chapter 3 we referred to a recent paper by Cheung concerning the proof of the inefficiencies associated with communal ownership.[1] His geometric proof of the dissipation of rent is shown by presupposing successive entries of new users of a scarce resource :

[1] S. N. S. Cheung, 'The structure of a contract and the theory of a non-exclusive resource', *Journal of Law and Economics*, vol. XIII, no. 4, 1970.

if there is only one user, then efficiency is attained, with two there is less efficiency, and so on, until the highest degree of inefficiency is reached with rent driven to zero when entry is complete and resource use has reached the saturation point. In the context of this simple model, each new entrant is assumed to be of the same size as each of the previous entrants, and it follows that the only way there can be any rent from the scarce resource is if entry is incomplete and resource use does not reach the saturation point. However, with free entry such a situation would not be attained except transitorily.

Throughout this study, we have had occasion at several instances to point out the simple historical fact that the commons in the open field villages were not available to anyone to use as he pleased. The rights of common were owned exclusively by a well defined collective of cultivators, and anyone who endeavored to use the commons without the consent of the village members committed legal trespass. In spite of this crucial difference from the Cheung model, it does not follow that his proof of rent dissipation under collective ownership is automatically inapplicable. The point is that unrestricted entry is only one mechanism that will result in rent being driven to zero. An alternative way to the same end is through over-use by those who have the rights of common. That is to say, we may show the same result by an amended version of the Cheung model, one in which the size of the enterprise is no longer constant but variable.

It follows that even if the commons were exclusive, this in no way guarantees efficient resource allocation. In the case of communal ownership exclusivity is only a necessary but not sufficient condition for efficiency – it is further necessary that each rights owner agrees to limit his use, in exchange for similar limitations on the use of others, so that over-use and rent dissipation do not result. We have seen that this was done in the open field villages by controlling or stinting the amount of livestock that each claimant of common rights was allowed to put to grazing. In this way the community ensured that rent was not dissipated, and this is the reason that the communal ownership of the commons was entirely consistent with private wealth maximization. We have endeavored to show in the preceding chapters how the gains from large-scale grazing could be attained more cheaply with collective rather than private ownership, and how these gains offset the costs of policing over-use. The conclusion is simply that Cheung's proof of the relative inefficiency of collective ownership is totally inapplicable to the open field system, for the preconditions of that model do not apply.[2]

It remains to be shown how the preceding discussion relates to the arguments presented by Demsetz, in the paper referred to above.[3] His claim to having established the relative efficiency of private over collective property rights also appears to fly in the face of the arguments presented here. Partly, however, the difference is semantic. In his discussion of the establishment of private ownership among the Montagnais fur hunters as a result of the growing scarcity with the advent of the fur trade, Demsetz makes the following statement :

By communal ownership, I shall mean a right which can be exercised by all members of the community. . . . Communal ownership means that the community denies to the state or to individual citizens the rights to interfere with any person's exercise of communally owned rights. Private ownership implies that the community recognizes the right of the owner to exclude others from exercising the owner's private rights. . . . under the communal property rights will take place mization of the value of communal property rights will take place without regard to many costs, because the owner of a communal right cannot exclude others from enjoying the fruits of his efforts and because negotiation costs are too high for all to agree jointly on optimal behavior. ,The development of private rights permits the owner to economize on the use of these resources from which he has the right to exclude others.

The problem is whether, on this definition of communal ownership, the commons of the open field system were communal or private property. The argument for using the phrase 'communally

---

[2] Cheung's analysis shows how scarcity is a sufficient condition for the inefficiency of communal access, and he simply asserts that it would not be feasible for the claimants to agree to and enforce restrictions of usership. However, this should be listed as a necessary *condition* for communal ownership to be inefficient. Thus, the *two* necessary and sufficient conditions for communal ownership to be inefficient are both scarcity and impossibility of enforcing restrictions on use. The point about the open field system is that it has a well established institutional mechanism for enforcing stinting rules, i.e., restrictions on the use of a common resource.

[3] H. Demsetz, 'Towards a theory of property right'.

owned' in the context is that all members of the relevant group, the open field village, had rights of common, i.e., partial ownership of the waste. The argument for using the phrase 'private ownership' is that no one outside this closed collective of cultivators had any rights whatsoever to the commons. They were private for a group, but communal for the members of that group. This is a case which Demsetz does not discuss, and his terminology is therefore not immediately applicable. Before we can establish an appropriate one it is necessary to dwell briefly on the case of the Montagnais reported by Demsetz.

It is easy to show that before the advent of the fur trade the rights of hunting were not just communal for some group of people, but that they in fact were much more general than that.[4] In fact, the Montagnais did not exclude anyone at all. Before the fur trade, different 'bands' seemed to have certain tracts of land on which they hunted. But the point in fact is that such boundaries were not respected at all. Anyone from any band could hunt anywhere he pleased, and only impracticalities of geographical divisions prevented them from doing so – the reason for the establishment of tracts for different bands. Furthermore, the bands were not closed groups. If someone wanted entry and membership in a new band, he was welcomed and would not be excluded. Animals were not scarce, so there were no economic problems. The case is one of a non-exclusive resource with insufficient entry so that some rent remained undissipated. It follows that what Demsetz has proved in his paper is the superiority of clearly defined rights in a scarce resource, as opposed to free usership. What does not follow is the inferiority of communal rights relative to private; for with communal rights we must understand rights as defined for a certain community, a well defined group of people obtaining certain well defined rights collectively as among themselves. Scarcity is a necessary condition for the delineation of rights: this is the proposition established by Demsetz. However, the rights so defined may be collective for a certain group, but not for outsiders of that group. It follows that economic theory, quite contrary to

---

[4] Demsetz, 'Towards a theory', pp. 354 and 356. The ensuing argument is based on an examination of the source referred to by Demsetz: E. Leacock, 'The Montagnais "Hunting Territory" and the fur trade', *American Anthropologist*, vol. 56, no. 5, part 2, memoir no. 78; especially pp. 6–7, 15, 20.

popular belief, does not necessarily imply the universal superiority of private rights over collective rights. What economic theory, correctly applied, does predict is that some type of rights, private or communal, will be defined when a resource becomes scarce. This gives the relationship between the theory presented here and the work by Demsetz – an extension and generalization of the basic insights provided in his paper. Collective ownership, if it is well defined, can be quite compatible with efficient resource allocation and private wealth maximization.

To cover this case, we really need to establish an entirely new category of ownership rights. We may think of them as 'collective exclusive' rights, or, to coin a new phrase, 'primunal' for something that is both private and communal. What has been shown in this essay is that there can be no presumption that there is a unique ranking in terms of efficiency of the three kinds of ownership rights that we now should recognize: private, collective exclusive or primunal, and collective. Whether one is more efficient than the other will depend (i) on the particulars of the technology of production, (ii) on the state of the market for the outputs and relative prices, since this determines the derived demand for productive resources, (iii) resource endowments and relative factor prices, insofar as this determines the choice of technology of production, and (iv) transaction costs, the usually forgotten element that turns out to be so crucial, for there is no unique ranking of transaction costs with respect to the three classes of ownership.

The theory of the firm

Lest it be thought that the analysis of a collective exclusive or primunal resource belongs only to a study of historical institutions, we shall proceed to show how the preceding analysis has some obvious applications in understanding the firm as an organization. It may already have been noted how striking are the resemblances between the organization of an open field village and of a modern-day corporation. Like the open field village, a corporation is an organization of joint owners of a scarce productive resource, capital. Just as the grazing areas of the village were owned collectively, so the capital is owned and controlled by a closed collective: no one owner owns a designated piece of machinery or buildings that he can call his own. Access to ownership in either of these organizations can only be achieved on prespecified rules : somebody must be induced to part with his membership of the organization, or additional resources brought in. The voting rule used in the two cases is virtually identical : the greater the share of the village lands or of the stocks in the firm, the greater the influence of each owner over the actions of the organization. The profit sharing rule is also the same, i.e., strict proportionality between inputs from each member and his share of the proceeds. It is remarkable how closely these two organizations parallel each other.

The firm is an institution where resource owners have joined together into a collective to realize certain economic gains from using their resources in common. If our conception of the institutional arrangements of the open field system is correct we should observe similar solutions to similar problems elsewhere. By pointing to the firm we are not only in a position to conduct a simple test of some of the implications of our theory of the open field system, but also perhaps to gain some insight into the firm as an institution.

The two seminal articles on the nature of the firm and the rationale for its existence are the ones by Coase, and Alchian and Demsetz. To repeat briefly, Coase's classic article views the firm as an organization that supersedes the market mechanism in those cases where there is a cost of transacting across markets. In the presence of such costs an institution that attains the proper allocation of resources and makes correct output decisions without transacting in markets will prove efficient. The Alchian–Demsetz theory of the firm is not necessarily inconsistent with Coase's. Rather, it builds on the foundations laid by Coase by attempting to identify exactly what costs the firm saves on. In their theory the firm serves to economize on certain policing costs that will arise in the presence of technological advantages of team production. On one very important point the two theories differ substantially : Alchian and Demsetz assert emphatically that the firm in no way represents a supersession of the market mechanism; instead, they assert that the firm constitutes a contractual arrangement arrived at by market exchange. The viewpoint taken here is slightly different; however, it will not be inconsistent with either Coase's path-breaking argument, or with Alchian and Demsetz' more specialized model.

H

In the case of the firm, as in the open field system, the necessary condition for the formation of an organization seems fulfilled : there are some elements of returns to scale that can be realized only by the grouping of individual agents. In the case of the corporation, the one we shall bring out here is the element of returns to scale in the use and procuration of capital : this is the reason several individual capital owners band together into a firm. Given that there are such returns to scale of grouping individual owners, there will be additional costs of decisionmaking for the group thus formed. Conceivably, if transactions were perfectly costless, such decisions could be formed by individual transactions by the parties involved. Each owner could deal individually with all the others. However, such transactions are costly in our world. A way to eliminate the need for them is to have all the individual owners agree on a joint decisionmaking formula. One of the most important costs of transacting that the firm as an organization thus serves to eliminate is the costs involved in collective decision-making in a group which jointly controls some resource. One further element is significant. Even with such a collective decision-making rule, there will be costs of having the collective make continuous decisions on minor aspects of the allocation of the jointly owned resource. In the open field system this was solved by appointing officers of the court : the reeve and the beadle, who served as overseers and had powers of decisionmaking with respect to the use of the communal resources between sessions of the court. This is the function of the manager in a modern-day corporation. He also serves to save on the costs for the members of the organization of making day-to-day decisions – he is the modern successor to the reeve and the beadle, as it were. Such a delegation of power follows naturally from the cost of communal decision-making.[5]

[5] The Alchian-Demsetz theory of the firm gives a nice explanation for why the owners of the firm, as team monitors, receive their payment in form of being residual claimants. What is provided here is a rationale for the establishing of a manager as team monitor : there are no team aspects to monitoring, so we should not expect to see the manager paid as a residual claimant. For a discussion of the historical accuracy of some of the background assumptions made by Demsetz, see J. C. McManus, 'An economic analysis of Indian behavior in the North American fur trade', *Journal of Economic History*, vol. XXXII, March 1972, pp. 36–53. The critique McManus levels against Demsetz does not affect the arguments presented here.

There is one further element of importance, which shows in another light the important parallel between the open field system and the modern corporation, and also shows the applicability of the third category of ownership rights analyzed in this paper: the capital stock of a typical firm is indeed nothing but a collective exclusive or primunal resource. What each capital owner has a claim to is an unidentified piece of real capital, and the total capital stock is owned jointly by the stockholders, just as each farmer in the open field village owns a piece of the commons that is only a right to a certain benefit stream. And, in addition, both the stockholder and the farmer can sell their claim, but cannot withdraw their piece, for they have no identifiable piece to withdraw. It serves the same purpose in both cases: for if there are elements of returns to scale in a firm or a corporation, then the bargaining power of a large owner can potentially enable him to extract the benefits accruing to other owners, if he has an identifiable piece of capital to withdraw. With joint ownership this problem is bypassed, as has been shown in the context of the open field system.

Thus, when all the members of the collective own only a share of the capital, and not any specified unit of that resource, the problem of policing the behavior of input owners is solved. In this context the fact that teams may be organized out of the employees of the firm is purely circumstantial. Such elements may well be present as complementary to the costs of decisionmaking among input owners stressed in the present context. The two formulations can together explain a vast array of phenomena: many firms have few owners, but important elements of team production; others have perhaps no elements of team production, but many owners. In such a firm there would be no problems of shirking and no costs of policing behavior, for there is no team production among the members of the organization. However, it is still a firm with all the characteristics that we normally associate with that organization.

The manner in which the open field system and the modern firm or corporation deal with transaction costs is also very similar. It was pointed out in Chapter 3 how we must interpret transaction costs due to information problems as being at least in part individual-specific. This means that some individuals may, simply because of their personal characteristics, be in a position to impose costs on others by their behavior. For example, in the open field system we pointed out how in various contexts it would be likely that some individual would be in a position to reduce the gains from cooperation to others, and thereby redistribute income or wealth to himself. In fact, one of the very important functions of the institutional organization of the typical open field village can be seen to be to deal with these issues. Specifically, both the scattering of the strips and the collective ownership of the commons were designed to cope with such problems. This is very similar to the problem faced in a modern firm or corporation, if we are to believe the Alchian–Demsetz theory of the firm. There it was pointed out how the benefits from team production can be reduced or dissipated by some individuals acting contrary to the interests of the team, and how this would necessitate a team monitor. This is a problem with only some individuals. It cannot be expected that all individuals would behave that way, or teams would become virtually impossible to organize. So both the firm and the open field village dealt with these individual-specific transaction costs in similar manner by setting up rules and incentive systems designed to minimize the likelihood of the organization becoming vulnerable to such influences.[6] It would seem that a crucial problem for organizations to deal with is to find viable ways of eliminating adverse behavior from individuals who participate in the organization.

There is one very important aspect in which the Alchian and Demsetz theory and our own version coincide. This is in the idea that the firm is an extension rather than a supersession of the price mechanism. Specifically, in this context we regard the firm as an organization for collective ownership established through a contractual arrangement consistent with private wealth maxi-

---

[6] In the open field system, team production was the order of the day; however, the problem of policing the team members into behaving properly was not accomplished by overseers and residual claimants. Instead, two alternative methods were resorted to: one was the scattering, which provided the team member with positive incentives to operate with team spirit; the other paying laborers partly in terms of assets belonging to the organizations, much like modern discussion of employee ownership of firms and corporations. This would also tend to put a premium on team spirit, and would reduce the problems of shirking.

mization. It is formed by resource owners joining their inputs together into a collective ownership by ceding to the firm their rights to make certain well specified decisions over the use of those resources. In exchange for this they receive an increase in their economic benefits from the resources thus joined. This is precisely the structure of a contract as it is generally thought of : a *quid pro quo*, with well specified terms, as decided on by two parties – in this case the firm and the resource owner. The firm is thereby created by a contractual arrangement whereby a bundle of property rights, i.e., some decisionmaking rights, are being transferred to the organization by its creators. Thus the firm cannot be regarded as superseding the market mechanism. On the contrary, it is the market mechanism that leads to its establishment.

Again, this view is entirely consistent with the model presented above of the open field system in medieval Europe. We may, if we wish, look at the open field village as a firm. It is a collection of decision rights created by a voluntary relinquishing of those rights by their owners. Implied in the relinquishing of those rights is a way of organizing the relative influence of each member of the collective thus created : a voting rule, and a way to share the proceeds, i.e., a profit sharing rule. As has been described above, such rules formed the very foundation of medieval agricultural society. What we have been able to show here is that this society is indeed entirely consistent with efficiency and private wealth maximization.

The firm as an organization is often thought of as the brainchild of the Industrial Revolution, and is considered by some a radical innovation. What is remarkable, when we consider the similarity between the firm and our interpretation of the open field village, is that the open field system was disappearing at the same time that the firm was making its appearance in the Industrial Revolution. The idea of the firm was not all that new after all. Its roots may now be traced back for as long as written records exist in open field villages, back to the early Middle Ages. There is one further fascinating little piece of evidence that strengthens this interpretation : in a very similar way to the firm, the members of the open field village were able to assume corporate responsibility and act as a juridical person. The village could enter into contractual agreements as one body, as for example in the renting of certain lands. It accepted joint responsibility in matters of taxation, militia, criminal liability, road and bridge servicing, and the like. As a body it could bind itself to fulfill obligations, and to incur financial liabilities. It could even sue and be sued as one body.[7] Thus even incorporation had its rudimentary form in the organizations of medieval agriculture. However, since the number of people involved in a typical village was usually very small – in the order of a few hundred, living side by side throughout their lives – formal legal arrangements sometimes took on a more informal shape. But the economic problems and their solutions were very modern.

Some observations on the nature of economic institutions

Our discussion of the similarities between the open field system and the modern corporation will enable us to point out some of the crucial elements in the construction of a more general theory of the formation and function of economic institutions. There would be little point in attempting to elaborate on a formal theory for the formation of economic institutions in this study; however, it would still appear incumbent on us to distill out of the foregoing discussion those basic principles about economic institutions that our inductive method will allow us to identify. These basic principles can then serve as initial guidelines for further work in this area.

First of all, we are now in a better position to specify the relationship between property rights and economic institutions in general. We showed in Chapter 3 how with the notion of property rights we ought to understand attenuated decisionmaking rights over the use and exchange of scarce assets. Furthermore, in our discussion of the open field system and its relevant functions we endeavored to show how the restrictions on decisionmaking rights that were so characteristic of the system can best be understood as having been established through a collective agreement that involved all the farmers with rights in the village. That is to say,

---

[7] For a very interesting account of some of these aspects, see Helen Maud Cam, 'The community of the vill', in F. L. Cheyett, *Lordship and Community in Medieval Europe*, New York, 1968, pp. 256–67.

we should describe the open field system as a collection of attenuated decisionmaking rights established through a process of mutually beneficial exchanges between self-interested economic agents. In principle, the case is very much the same with respect to the firm. Here, the capital owners have voluntarily relinquished certain decisionmaking rights over the scarce capital assets that they control, and this is what is at the heart of what constitutes the relationship we call the firm.

What therefore defines the institutions, the open field village and the modern firm, is that they are empowered to make certain decisions, and that they have in a specific manner acquired those rights from their original owners. Insofar as economic institutions are independent decisionmaking agents, in the way that the organizational bodies of the open field village and the firm are in a position to make decisions over the use of scarce economic assets, it is clear that what constitutes their ability to make these decisions is that a conditional transfer of decisionmaking powers has been made to them. The decisions made by the manorial court or the open field village meeting were viable because the constituency that had initially formed that decisionmaking body also recognized and abided by the decisions handed down by the court or the meeting. What made the court able to wield decisionmaking power was simply the recognition of that power by the interested parties, the very parties that by this recognition also vested the decisionmaking body with property rights. The case is the same with the firm. The power to make decisions over the use of the capital equipment is conditionally granted to the firm by the original owners of the capital. It is the agreement to abide by the authority of the firm that lays the foundation for the firm as a viable decisionmaking organization.

However, there are other kinds of economic institutions besides those that, like the firm and the open field village, actually constitute active decisionmaking bodies. We often refer to customs, traditions, social codes and mores as institutions as well, although it is clear that there is no man-made creation of a new decisionmaker in those cases. Can we fit such informal institutional considerations into the present terminology as well? To see this, we should inquire into what the real function of these informal institutions is. In essence, they serve to make behavior more predictable by imposing restrictions on individuals in society. They do this not by continuous decisionmaking, but rather by inflicting social sanctions on those who do not conform to generally accepted modes of behavior. The more traditional and conventional a society, the stronger those sanctions will be. It is to be noted that the result is a limitation of freedom for the individual : to remain accepted, he must behave as others do. This limits his decision-making scope, attenuates his property rights as it were, and simultaneously serves to make his actions more predictable for the rest of society.

In this manner we can clarify the link between institutions and the economic concept of property rights. Economic institutions are really represented by a collection of property rights or decision-making powers that are attenuated in various specific ways. We may make a distinction, if we wish, between those institutions that make independent decisions and those that are of a general nature and restrict individual behavior in the way social conventions do. However, the important key to an understanding of any such social or economic institution is that it remains viable only through attaining a measure of social consent to its functions, i.e., its scope of influence is really ultimately dependent on voluntary agreement and cooperation by the economic agents whose decisionmaking powers are limited by the existence of the institutions.

Referring back to Chapter 3 once more, we may now also consider the distinction between institutions that are instituted by either the government, the supreme level of public legislation, and the courts – by decisionmakers that are outside the immediate purview of the market mechanism – or by those that are the result of market-like arrangements, such as the ones we have considered in this study. Although it is perfectly possible to regard the former as also resulting from implicit trading, it is not clear that we can usefully apply the same efficiency criteria we have shown to be relevant to the open field village or the modern corporation. At this stage of analysis, it is desirable to limit the discussion to institutions that are immediately created through voluntary contracts within the market mechanism.

We may then regard these endogenously created institutions as set up by trading parties who exchange decisionmaking rights for an improved real income, i.e., a better use of scarce resources. This is attained through the avoidance or decrease of costs of transacting that otherwise would consume real resources. Such costs of transacting are often associated with adverse behavior by some individual or individuals. We may look at the behavior of all individuals as being classifiable into various categories on a scale from extremely cooperative to extremely adverse. Whether that distribution is normal, uniform, binomial, or of some other form is irrelevant for the argument. The function of institutions is to find ways and means by which (i) the negative results imposed on others by those in the adverse tail of the distribution are minimized, and (ii) the incentive for persons to adopt behavior as exemplified by those in the adverse tail is minimized. This is a view that derives directly from our study of the firm, the open field system, and the nature of transaction costs as being individual-specific. We have discussed how various institutional arrangements perform this function in different manners, and how their relative efficiency must be ranked with respect to how changes in property rights or decisionmaking rights that result from the formation of some institutions affect the constraints under which individuals perform their constrained optimization.

By eliminating the negative or adverse behavior in one tail of the distribution of individual behavior, it is clear that the economic effect of institutions is to make the behavior of the remaining individuals more predictable. In this manner contracts that might otherwise have been impossible to enter into may become feasible, and so trades that are mutually beneficial for the involved parties become established. The net result is an increase in real income for society as a whole.

However, it is crucial to the analysis to keep in mind the twofold function of economic institutions as they involve the phenomenon of property rights. We discussed in Chapter 3 the role of property rights as determining income or wealth distribution, on the one hand, and also serving as signals for behavior, thereby guiding incentives, on the other. If we accept the view that institutions are really nothing but specific collections of attenuated property rights, then it follows that institutions are also tied up with both income distribution and incentive formation. This generates two kinds of complications. The first is that the question of the relative efficiency of various institutions now becomes dependent on our ability to rank various income distributions. For if every set of institutions is associated with a certain distribution of income, and if we wish to ascertain which set of institutions is the most efficient with respect to transaction costs, then we are also, implicitly or explicitly, comparing various distributions of benefits associated with particular institutions. This casts rather grave doubts on the question whether there ever will be a purely economic theory of institutions : for the science of economics has, along with every other discipline, been unable to determine rules for comparing the desirability of different distributions of income or wealth. This is an issue inextricably tied up with ethical judgments, moral considerations, and questions of personal and political value systems, and so outside the purview of science proper. If this is the case, then it will also follow that a full analysis of institutional efficiency is a problem more for the polity than for academics.

The second complication that arises from the realization that institutions, through the link of property rights, are tied in with income distribution is that any analysis of institutional change will be exceedingly complex. For if institutions are associated with certain distributions of benefits, then a change of institutional structure will imply a change in the distribution of those benefits, except in those few cases when new institutions also yield an increase in income that is sufficient to avoid any effective redistributions. This provides one clue to why institutional change can be expected to be both traumatic and discrete rather than continuous. It will be traumatic because those who stand to lose from the change to new institutions will often not give in easily. It will be discrete rather than continuous because it will come about only if the gains from the institutional change are large enough to overcome resistance from those who might lose from a change and in addition increase real income. These are transaction costs considerations, the relevance of which we have already shown in the context of the enclosure of open fields. The implication is that it will be very difficult to formulate conditions for when, from a purely conceptual standpoint, a change of institutions is efficient or not. It is clear that adverse behavior by some may be severe enough, at least in some instances, to eliminate any gain from changing to a new kind of organization. Potentially there may be a loss of real income to society from such behavior, so it would appear to be inefficient not to change. However, if we are guided by actual behavior as yielding information about the relative strength of various individuals' preferences, then we might conclude that the fact that some people frantically attempt to avoid an institutional change may be reason enough to not undertake that change. This is a question of what the proper point of reference for judgements about efficiency really is; a complex question of importance for institutions and institutional change, but quite outside the field for analysis in this context.

There is one further reason for believing that institutional change in many instances will be of a rather drastic nature. This has to do with the interrelations between technological change and existing institutions. It was argued in the context of the open field system that that system may have been prone to producing certain technical solutions to its production problems that suited its institutional environment. To the degree that such new techniques were successfully implemented, it is clear that technological change did not topple the open field system, but instead served to reinforce it as a viable organization. That is to say, certain directions of technical change may make further institutional change more difficult by making existing institutions more vigorous. However, there might be some drastic changes in technology, perhaps discovered by a piece of fortuitous insight, that could only be implemented through a new institutional setting. Such changes would probably be rather sweeping and drastic.

The conclusion of this is that, even if we realize that there is a strong element of economic efficiency in the functioning of various economic institutions, there are several random elements that will effectively make it impossible for us ever to construct a theory of institutions and institutional change that relies exclusively on standard choice theory and its optimality and efficiency theorems. These random elements are associated partly with transaction costs and the individual-specific component of such costs, partly with the nature of technological progress that can never be completely foreseen and the effects this has for judging the perform-

ance of existing institutions. Until we get a deeper insight into
these random factors in institutional evolutions, we shall have to
avoid any far-ranging generalizations of a choice theory of endogenous economic institutions. Until such a time, it will suffice to be
specific. We can still analyze historical institutions, and compare
them with the performance of known alternatives – this has been
the method of our analysis of the open field system. In a sense,
the most important generalization to come out of the attempt to
establish a purely economic theory of the open field system is,
therefore, a negative one. It is quite likely that this theory,
although useful in the specific context, cannot be carried much
further but must remain confined to specific situations where we
can identify various given alternatives and compare them with
each other. Thus the method would seem to be useful and important, but its applicability limited to specific institutions and
instances.

The property rights approach to history

We have shown how the property rights and transaction costs
approach has enabled us to get a fresh perspective on the open
field system, and how this carries over into a more general analysis
of institutions and institutional change. These are issues that have
a long tradition in economics as well as history. It may, therefore,
be appropriate to attempt to specify the relationship between the
property rights approach and more traditional methods of analyzing historical and economic problems.

   At the heart of any scientific inquiry is what may be called the
colligation problem, the problem of determining what can safely
be left outside the analysis and what must be retained to ensure
a proper understanding of the issues. In economic terminology,
this is simply a determination of what is to be exogenous and
endogenous in the particular model to be constructed. What
makes the property rights approach so attractive is partly that it
solves this colligation problem in a different manner from the
more traditional approach adopted in historical analysis. We shall
make some simple observations on the relationship between this
approach and standard economics, Marxian analysis, and traditional historical research, respectively.

In Chapter 3, we observed that the property rights approach may be said to interject an intermediate step into a Walrasian paradigm, and this gives the key to an evaluation of the approach in relation to standard economics. Since the rediscovery by Hicks in the 1930s of Walras' general equilibrium approach, this has become the dominant method in modern economics. Even though pockets of non-Walrasian economics remain, we still have a firm enough grasp of the relationship between, for example, Marshallian and Walrasian[8] theory for us to be content to use Walrasian analysis as the numeraire for economics in general. Thus many, or perhaps even most, existing propositions of traditional economics will continue to apply to property rights theory.

The intermediate step inserted by the property rights approach is explicitly to make endogenous a set of relations formally left out in standard contemporary economics : the allocation of property rights and, by implication, the making of economic institutions. It is curious how the dominant paradigm of contemporary economics is completely void of institutions, and how, as a consequence, it is totally unsuited for application on its own to historical research. It seems clear that one reason for this lack of interest in institutions in Walrasian economics is that the paradigm, by its very design, is unable to deal with transaction costs in any reasonable manner.[9] In order to prove existence, uniqueness, stability, and Pareto optimality of the competitive equilibrium, the Walrasian approach must rely on a whole host of artificial assumptions – most of which are explicitly designed to ensure that transaction costs, as conceived of throughout this essay, are kept at zero. The result is that institutions become irrelevant : we have shown how it is possible to understand the economic function of institutions only by relating them explicitly to the economic consequences of transaction costs.

The price that must be paid for putting transaction costs in the constraints and making property rights and institutions endogenous choice variables is that at least some of the widely accepted results of economic theory can no longer automatically be

---

[8] For a further discussion of the issues involved in this statement, see Axel Leijonhufvud, 'Varieties of price theory', unpublished manuscript.

[9] The only exceptions are the uninteresting fixed and well known set-up and transfer costs analyzed in Chapter 3.

presumed defensible. We have hinted on several occasions earlier that this will include the notion of efficiency in accepted economic doctrine. With efficiency is normally understood the Pareto optimum of the competitive equilibrium. However, if we accept the definition of transaction costs proposed in this essay, as well as the realization that transaction costs are individual-specific, rather than market-general, then we must conclude that in the presence of transaction costs there is no such thing as a unique competitive equilibrium, for any attainable competitive equilibrium will be a function of the particular individuals involved in the economic game; their idiosyncrasies cannot be assumed away if transaction costs are to be included in the analysis. As a consequence, we shall have to accept that the competitive equilibrium of standard analysis, i.e., with no transaction costs, will not be a Pareto optimum when transaction costs are included in the analysis. This means that any propositions of welfare theory, insofar as they rely on the competitive equilibrium as a point of reference, become totally untenable in property rights analysis.[10] As we shall see, however, this affords the application of property rights analysis to history some decisive advantages from a conceptual viewpoint.

Yet even if we have to reject the notion of global efficiency when we include transaction costs in the analysis, we shall still be able to use a more limited notion of efficiency. This is the approach taken in standard marginal analysis, in the Marshallian tradition. What is inherent in the marginalist approach is a comparison, on the margin,[11] of existing solutions with a well specified alternative – just as we have compared the solutions of the open field system with those of enclosure as a relevant alternative. As a consequence, we shall be able to retain the notion of Pareto optimality, for any proposed change can always be compared with the presently existing solution, even if individual-specific transaction costs are realized to be important. Furthermore, most of the accepted propositions of production and consumption theory will remain applicable, but will have to be interpreted in the light of imperfect information and costly transacting. Thus

---

[10] These issues are addressed further in Carl J. Dahlman, 'The problem of externality', *Journal of Law and Economics*, vol. XXII, 1979.

[11] The use of the word margin here should not be confused with a derivative. The changes contemplated may well be discreet and far from infinitesimal.

in some respects, property rights theory is an important addition to standard economic theory.

Another couple of elements that are missing in received economic doctrine as applied to historical problems are supplied by Marxian analysis. This is the stress on income distribution problems, on one hand, and on political mechanisms, on the other. At the heart of the Marxian critique of 'neoclassical' economic analysis, whatever that is taken to mean,[12] is its neglect of uses and abuses of power as a means of attaining skewed distributions of economic influence and income. Perhaps it is the failure of standard economic theory to address these issues successfully that accounts for the continued presence of Marxian doctrines in contemporary historical, as well as economic, inquiry. However unsuccessful Marxian doctrine itself has been in contributing a positive analysis of such questions, it must be admitted that until recently it remained the only avenue for even attempting to deal with these issues in any systematic manner, whether the context be historical or otherwise.

Yet the failures and limitations of Marxian analysis are severe restrictions on its use as a tool of historical research. Perhaps the most important failure is the inability of Marxists to formulate empirically testable and falsifiable implications.[13] We endeavoured to show in the critical evaluation of the Marxist approach to the open field system in Chapter 2 that the Marxist model of the system always seems to be consistent with any observed events, and cannot therefore be considered anything but a formal tautology, void of serious analytical content. Ultimately, this must be considered due to the moral and ethical content of Marxist ideology. The drawback of the Marxist approach to history is that it inevitably prejudges and presupposes what ought, in his-

---

[12] 'Neoclassical' is one of the fuzziest phrases in modern economics. Neoclassical theory is not a school of thought, not a set of questions, not a set of answers, not a methodology. Even Marxian analysis implies constrained optimization procedures, with different constraints and objective functions, to be sure. Usually, it seems that 'neoclassical' is a label pinned on a member of an opposed line of reasoning – and therefore best to be avoided completely.

[13] However, some Marxian writers contend that it is not a purpose of Marxian analysis to establish empirically refutable propositions about history. See e.g., B. Hindess and P. Q. Hirst, *Pre-capitalist Modes of Production*, London, 1975.

torical analysis, to be left as a derived and proven conclusion : who the villains of history were. In the Marxian analysis of class contradictions as the engine which moves history forwards there are essentially only two opposing forces : those who own the means of production, and who use these means for the exploitation of those who do not own. The purpose of this dichotomous class structure would simply seem to be to establish clearly the simple policy conclusion : take from the haves, who abuse their power, and give it to the have-nots, who justly have the right to what has been taken from them.

In the preceding, we have been able to show how the property rights approach enables us to deal with these and similar issues of a traditional Marxian brand in a manner that bypasses these three drawbacks. First, we noted how it is impossible to analyze property rights without at the same time recognizing the fact that property rights constitute the social means of allocating rights to income, as well as political and economic decisionmaking power. Thus income distribution problems are at the very heart of the property rights approach.[14] Secondly, the property rights approach leaves the possibility open that various groups can form explicit or implicit coalitions, or classes, by voluntary agreement, and also recognizes that even the losers in institutional change possess bargaining power that they can use to their advantage – a proposition we considered in the context of enclosure, pointing out that the fact that many complaints were lodged against enclosure cannot be taken to mean that those who complained also lost. Complaints may be a way of influencing the process and a means of imposing costs on proponents of change for the benefit of those who complain. In this manner, the property rights approach will be able to deal with a greater number of classes than two, and thereby relaxes a very restrictive assumption of classical Marxian analysis. Thirdly, the property rights approach takes no moral or ethical side in prejudging the normative aspects of any historical situation – that is an issue left open for anyone to determine in

---

[14] For example, in this essay, we have ventured to show how the collective ownership of the commons as well as the scattering of the strips in the arable served to protect the small and the weak from unwanted income redistributions to the rich and the powerful. Thus, at the very heart of the structure of the open field village was an attempt to deal with a tricky income distribution problem through the determination of property rights.

the way he pleases. It would therefore seem that anything a Marxist analysis of history can do, property rights and transaction costs can do better – especially since it allows most of the standard tools of received economic doctrine to remain applicable, and hence allows for the possibility that voluntary transactions on mutually agreeable terms actually exist.

On the other hand, the implications of properly applied property rights analysis for historical research would seem to be very traditional in some very crucial respects. We pointed out above that the realization that transaction costs must be indexed over individuals, not over markets or goods, implies that the global efficiency of the competitive equilibrium can no longer be considered relevant as a Pareto optimum. In plain English, this means that there is absolutely no analytical reason why we should presume two countries with similar resources endowments, identical technologies of production, and same preferences to have a historical process that turns out identical products. The reason is that various random elements in the make-up of the historical context, such as for example the influence of particular individuals, will inevitably make the two societies pass through different evolutionary schemes since the structure of transaction costs will differ between them. Thus the property rights approach completely bypasses the oft-repeated criticism levelled by traditional historians against the application of economic analysis to history : the apparent implication of determinism. When we introduce individual-specific transaction costs as an element of the analysis, solutions will no longer be unique, and there can consequently be no determinism.

This means that, if we are to gain an understanding of any historical process or phenomena, we shall have to marry the economist's approach in analyzing the tendencies towards economic efficiency in the allocation of resources with the Marxist's insistence that such tendencies may be thwarted by self-interested parties who use the structure to their own advantage, and all the time pay heed to the traditional historical tenet that it is only by a close study of the elements of the historical context itself that we shall learn the truth. It would seem that the property rights approach combines the best aspects of these three methods, while at the same time opening up new questions and yielding new answers to old ones. The belief of this writer is that a systematic exploration of the implications of property rights analysis, as well as its consistent application to historical problems, will be able to do much to break down the present artificial barriers between history, economics, and political science, and between Marxists and 'neoclassicals'. It is a rich and promising approach; yet it is likely that established interests in the form of sunk costs invested in the acquiring of firm beliefs and doctrinal divisions between various disciplines will considerably delay the introduction of a unified methodology of historical research – in itself a prediction quite consistent with a transaction costs and property rights view of the world.

BIBLIOGRAPHY

Alchian, A. A., and Demsetz, H., 'Production, information costs, and economic organization', *American Economic Review*, vol. 43, December 1972, pp. 777–95
and Kessel, R., 'Competition, monopoly, and the pursuit of pecuniary gain', in *Aspects of Labor Economics*, National Bureau of Economic Research, Princeton, 1962
Allison, K. J., 'Flock management in the sixteenth and seventeenth centuries', *Economic History Review*, 2nd series, vol. XI, 1958–9
Ault, W. O., *Open-Field Husbandry and the Village Community*, Transactions of the American Philosophical Society, Philadelphia, 1965
Open-field Farming in Medieval England, London, 1972
Baack, B. O., and Thomas, R. P., 'The enclosure movement and the supply of labor during the Industrial Revolution', *Journal of European Economic History*, vol. III, 1972, p. 401 *et seq.*
Baker, A. R. H., and Butlin, R. A. (eds.), *Studies of Field Systems in the British Isles*, Cambridge, 1973
'Field systems of southeast England', in Baker and Butlin (eds.)
Bishop, T. A. M., 'Assarting and the growth of the open fields', *The Economic History Review*, vol. VI, 1935–6, pp. 13–29
Bloch, M., 'The rise of dependent cultivation and seignorial institutions', in Postan (ed.)
Blum, J., 'The internal structure and polity of the European village community from the fifteenth to the nineteenth centuries', *Journal of Modern History*, vol. 43, 1971, pp. 549–52
Bowden, P. J., *The Wool Trade of Tudor and Stuart England*, London, 1962
Bracton, H. de, *Laws and Customs of England*, Cambridge, Mass., 1977
Bradley, H., 'The enclosure in England : an economic reconstruction', *Columbia University Studies in History, Economics, and Public Law*, vol. LXXX, no. 186, 1918
Brooke, C. N. L., and Postan, M. M. (eds.), *Carte Nativorum*, Oxford, 1960

Buchanan, J. M., *The Demand and Supply of Public Goods*, Chicago, 1968

and Tullock, G., *The Calculus of Consent*, Michigan, 1962

Cam, H. M., 'The community of the vill', in Cheyett (ed.)

Carrier, E. H., *The Pastoral Heritage of Britain*, London, 1936

Carus-Wilson, E. M. (ed.), *Essays in Economic History*, vol. I, London, 1966

'The woollen industry before 1556', in Pugh (ed.)

Chambers, J. D., 'Enclosure and the small landowner', *Economic History Review*, vol. IX, 1938–9, pp. 118–27

'Enclosure and labour supply in the Industrial Revolution', *The Economic History Review*, 2nd series, vol. V, 1952–3, pp. 319–43

Nottinghamshire in the Eighteenth Century, London, 1966

and Mingay, G. E., *The Agricultural Revolution 1750–1880*, London, 1966

Cheung, S. N. S., 'The structure of a contract and the theory of a non-exclusive resource', *Journal of Law and Economics*, vol. XIII, April 1970, pp. 49–70

Cheyett, F. L., *Lordship and Community in Medieval Europe*, New York, 1968

Clower, R. W., 'Foundations of monetary theory', in Clower, (ed.) (ed.), *Monetary Theory*, Harmondsworth, 1969

Coase, R. H., 'The nature of the firm', *Economica*, vol. IV, 1937, see pp. 125, 386–405

'The problem of social cost', *Journal of Law and Economics*, vol. III, 1960, pp. 1–60

Cohen, J. S., and Weitzman, M. L., 'A Marxian model of enclosures', *Journal of Development Economics*, vol. I, no. 4, February 1975, pp. 287–336

Curtler, W. H. R., *The Enclosure and Redistribution of Our Land*, Oxford, 1920

Dahlman, C. J., 'The problem of externality', *Journal of Law and Economics*, vol. XXII, 1979

Demsetz, H., 'Towards a theory of property rights', *American Economic Review*, vol. LVII, May 1967, pp. 347–59

'Information and efficiency : another viewpoint', *Journal of Law and Economics*, vol. XII, 1969, p. 1 *et seq.*

Dobb, M., *Studies in the Development of Capitalism*, London, 1946

Dyos, H. J., and Aldcroft, D. H., *British Transport*, Leicester, 1969

Elliott, G., 'Field systems of northwest England', in Baker and Butlin (eds.)

Lord Ernle, *English Farming, Past and Present*, London, 1968

Everitt, A., 'The marketing of agricultural produce', in Thirsk (ed.)

Fisher, F. J., 'Commercial trends and policy in sixteenth century England', *Economic History Review*, vol. X, 1940

Flinn, M. W., *Origins of the Industrial Revolution*, London, 1966

Furubotn, E. G., and Pejovich, S., 'Property rights and economic theory : a survey of recent literature', *Journal of Economic Literature*, vol. X, no. 4, December 1972, pp. 1137–62

(eds.), *The Economics of Property Rights*, Cambridge, 1974

Gay, E. F., 'Inclosures in England in the sixteenth century', *Quarterly Journal of Economics*, vol. XVII, 1903, pp. 576–97

Gonner, E. C. K., *Common Land and Inclosure*, London, 2nd ed., 1966

Gray, H. L., *English Field Systems*, Cambridge, Mass., 1959

Griggs, D., *The Agricultural Revolution in South Lincolnshire*, Cambridge, 1966

Hammond, J. L., and Hammond, B., *The Village Labourer*, New York, 1965

Havinden, M. A., 'Agricultural progress in open-field Oxfordshire', in Minchinton (ed.)

Hilton, R. H., *A Medieval Society: The West Midlands at the End of the Thirteenth Century*, London, 1966

Hindess, B., and Hirst, P. Q., *Pre-capitalist Modes of Production*, London, 1975

Hoffman, R. C., 'Medieval origins of the common fields', in Parker and Jones (ed.)

Homans, G. C., *English Villagers of the Thirteenth Century*, New York, 1970

Hoskins, W. G., *Provincial England*, London, 1963
The Midland Peasant, London, 1965
and Stamp, L. D., *The Common Lands of England and Wales*, London, 1963

John, A. H., 'The course of agricultural change', in Pressnell (ed.)
'Farming in wartime 1793–1815', in Jones and Mingay (eds.)

Jones, E. L. (ed.), *Agriculture and Economic Growth in England 1650–1815*, London, 1967
and Mingay, G. E. (eds.), *Land, Labour, and Population in the Industrial Revolution*, London, 1967

Kerridge, E., *The Agricultural Revolution*, London, 1967
Agrarian Problems in the Sixteenth Century and After, London, 1969
'Agriculture 1500–1973', in Pugh (ed.)

Klein, B., 'Competitive interest payments on bank deposits', *American Economic Review*, vol. LXIV, 1974

Laslett, P., *The World We Have Lost*, New York, 2nd ed., 1971

Lavrovsky, V. M., *Ogorazhivanye Obshchinykh Zemel v Anglii*, Moscow, 1940

Leacock, E., 'The Montagnais "hunting territory" and the fur trade', *American Anthropologist*, vol. 56, no. 5, part 2

Leijonhufvud, A., 'Varieties of price theory', Working Paper no. 40, 1974, Department of Economics, UCLA

McCloskey, D. N., 'English open fields as behavior towards risk', in Uselding (ed.)
'The persistence of English common fields', in Parker and Jones (eds.)
'The economics of enclosure', in Parker and Jones (eds.)

McManus, J. C., 'An economic analysis of Indian behavior in the North American fur trade', *Journal of Economic History*, vol. XXXII, March 1972

Mantoux, P., *The Industrial Revolution in the Eighteenth Century*, London, 1947

Martin, J. M., 'The parliamentary enclosure movement and rural society in Warwickshire', *Agricultural History Review*, vol. XV, 1967

Marx, K., *Capital*, Chicago, 1906, vol. III

Minchinton, W. E. (ed.), *Essays in Agrarian History*, Newton Abbot, England, 1968

Mingay, G. E., *English Landed Society in the Eighteenth Century*, London, 1963

North, D. C., and Thomas, R. P., *The Rise of the Western World*, Cambridge, 1973

Orwin, C. S., and Orwin, C. S., *The Open Fields*, Oxford, 1967

Parker, W. N., and Jones, E. L., *European Peasants and Their Markets*, Princeton, 1975

Pollock, F., and Maitland, F. W., *The History of English Law*, Cambridge, 1968

Postan, M. M. (ed.), *The Cambridge Economic History of Europe*, Cambridge, 2nd ed., 1966
The Medieval Economy and Society, Berkeley and Los Angeles, 1972
and Power, E., *Studies in the English Trade in the Fifteenth Century*, New York, 1966

Pressnell, L. S. (ed.), *Studies in the Industrial Revolution*, Oxford, 1960

Pugh, R. (ed.), *A History of Wiltshire* (Victoria History of the Counties of England), Oxford, 1960

Raftis, J. A., *Tenure and Mobility*, Toronto, 1964

Ramsey, P., *Tudor Economic Problems*, London, 1965

Reid, J., 'Sharecropping as an understandable market response : the post-bellum south', *Journal of Economic History*, vol. XXXIII, 1973

Roberts, B. K., 'Field systems of the west midlands', in Baker and Butlin (eds.)

Ruffhead, O., *Statutes at Large*, London, 1743

Schumpeter, J. A., *History of Economic Analysis*, Oxford, 1954

Seebohm, F., *The English Village Community*, London, 1912

Slater, G., *The English Peasantry and the Enclosure of Common Fields*, New York, 1968

Smith, R. B., *Land and Politics in the England of Henry VIII*, Oxford, 1970

Stigler, G. E., and Boulding, K. E. (eds.), *Readings in Price Theory*, Irwin, 1953

Tate, W. E., *The English Village Community and the Enclosure Movements*, London, 1967

Tawney, R. H., *The Agrarian Problem of the Sixteenth Century*, London, 1912

Thirsk, J., *Tudor Enclosures*, Historical Association Pamphlet no. 41, London, 1959

(ed.), *The Agrarian History of England and Wales*, Cambridge, 1967

'The common fields', *Past and Present*, no. 29, 1964, pp. 3–25

'Field systems of the east midlands', in Baker and Butlin (eds.)

Tillinghast, P. E., *The Specious Past*, Reading, Mass., 1972

Titow, J. Z., 'Medieval England and the open-field system', *Past and Present*, no. 32, 1965

English Rural Society, 1200–1350, London, 1969

Turner, M. E., 'The cost of parliamentary enclosure in Buckinghamshire', *Agricultural History Review*, vol. XXI, 1973, pp. 35–46

'Parliamentary enclosure and landownership change in Buckinghamshire', *Economic History Review*, 2nd series, vol. XXVIII, no. 4, 1975

Uselding, P. (ed.), *Research in Economic History: An Annual Compilation*, vol. I, 1976

van der Wee, *The Growth of the Antwerp Market and the European Economy*, The Hague, 1963

Vinogradoff, P., *Villainage in England*, Oxford, 1892

Yelling, J. A., *Common Field and Enclosure in England, 1450–1850*, London, 1977

Yonekawa, S., 'Champion and woodland Norfolk : the development of regional differences', *Journal of European Economic History*, vol. 6, no. 1, 1977

INDEX OF NAMES

INDEX OF SUBJECTS